TEACHING WITH AI:
The PEACE Framework

A Practical Guide to AI-Powered Inquiry-Based Learning for 21st-Century Skills

AYO JONES, M.Ed.

TEACHING WITH AI: The PEACE Framework

Noodle Knowledge Publishing
Knowledge for Your Noodle
© 2024 Ayodele Jones

For permissions, email hello@noodleknowledge.com

Or mail to;
Noodle Knowledge Publishing
PO Box 112
Oregon, WI 53575

Printed in the United States of America.

ISBN: 979-8-218-56826-9

AUTHOR'S NOTE

Is this book for you?

Let me take you back to a time when chalkboards ruled the classroom. The technology revolution was rolling in the TV cart on loan from the librarian. We've come a long way since then, haven't we?

AI is changing everything about how we live, work, and play. No, this is not a warrior cry to rise up against the machines. Quite the opposite, actually. Instead of obsessing about how kids can use chatbots to cheat, let's focus on how AI can truly revolutionize how we learn. After all, despite astronomical leaps in technology since the chalkboard days, there are still teachers ignoring AI altogether or just using it like a fancy worksheet factory? I mean, really? Are we going to let this incredible tool be overlooked or reduced to a glorified quiz maker when it could be so much more? It's a huge missed opportunity.

In order to meaningfully adopt AI tools into the classroom, teachers have to learn some new tricks. First, educators must learn about AI tools, prompt engineering, and tech integration. Using technology in the classroom is a little more complicated than it was when we rolled in that TV cart. Second, and most important, we must stop focusing on memorization or standardization and start helping students apply their learning through creation and innovation with the support of AI tools.

It's a huge leap. I know this. And when we take huge leaps and work to reframe our approach to education, it's only natural to encounter some bumps along the way. While the benefits of AI are exciting, fears about cheating, replacing teachers, and student dependency can create some serious roadblocks.

This book is about reframing instruction by embracing AI and pivoting towards a focus on developing skills, including critical thinking, problem solving, communication, collaboration, and prompt engineering in students using the PEACE Framework. It's a practical approach to AI-powered inquiry-based learning to help you guide your student through meaningful explorations.

AI could be creating real, personalized learning experiences, allowing students to ask questions that go way deeper than Googling for information. I invite you to join me in exploring how AI can be used to build a future where our students aren't just passive AI users - they're creators, innovators, and leaders using AI to expand what's possible.

AYO JONES, M.Ed.

The PEACE Framework

P.S. This book was not written by AI, it was written by Ayo (that's me). Yes, I absolutely used LLMs to brainstorm, flush out thoughts, edit text, and reword for clarity, but the ideas, structures, and frameworks are my own. As you use technology to reiterate and reimagine educational systems and instructional pedagogy through the lens of what I've outlined in this book, I just ask that you link back to the original author, credit where credit is due, and refrain from plagiarism, theft, or trickery.

TABLE OF CONTENTS

Introduction

In the blink of an eye, artificial intelligence (AI) has burst onto the scene, transforming everything from the way we shop to the way we communicate. It seems like just yesterday we were marveling at Siri and Alexa's ability to understand our voice commands; now, AI is writing poetry, composing music, and even outsmarting us at our own games.

With all this rapid advancement, we have to ask: **Are our schools keeping up?** Are we preparing students for a future where AI is not just a tool but an integral part of their daily lives? This question is especially critical within K-12 education as we shape young minds and prepare students for life in a new AI-powered world.

So, is education keeping up with AI? The short answer is: "Not even close."

Many schools and districts worldwide limited access to AI tools and large language models (LLMs) early on. They erred on the side of caution, fearing how AI might negatively impact students, worrying about AI being used for cheating, and adopting a rather closed mindset. Overwhelmed teachers saw the capacity of

AI tools to ease their workload and secretly chatted with LLMs to get help with lesson planning, curriculum modifications, and even data analysis. However, when it came to student interactions, teachers often continued with traditional instructional methods because assessments and lessons hadn't adapted to account for the presence of AI.

At this moment, where we see monumental leaps in the capacity of technology every day, why are so many teachers not using generative large LLMs and why are the majority of classrooms still untouched by AI? It's a huge missed opportunity. Just think about what's possible: AI could be creating real, personalized learning experiences, allowing students to ask questions that go way deeper than Googling for information. Instead of focusing on memorization, AI could help students apply their learning through creation and innovation. Picture students using LLMs to collaborate on creative projects, dig deep into research, and sharpen their critical thinking. These are just a few ways AI could focus on inquiry and get students ready for the challenges and opportunities of the 21st century.

In this book, we're going to dig into how AI can truly revolutionize education. Not by replacing teachers (don't worry, we're not going there) but by giving them the tools to teach with an approach we've always known was effective, but struggled to implement. We'll look at how AI can spark curiosity, boost critical thinking, and actually prepare students for landing a job and "adulting" in the real world.

Ready to shake up your traditional teaching methods and embrace the transformative power of AI? Let's figure out how AI can help us build a future where our students aren't just passive AI users - they're creators, innovators, and leaders using AI to expand what's possible.

Chapter 1

Asking Better Questions: The Next Frontier

As a kid, I remember huddling around the TV with my family watching Jeopardy.

Am I the only one?

I was always amazed at how the contestants seemed to be like walking encyclopedias. Their minds were just these vast repertoires of knowledge and facts. I literally marveled at their ability to recall obscure ideas and answer seemingly impossible questions.

That was actually how we all operated. We were really only as smart as we were capable of remembering vast amounts of information. When I was a kid, I could remember the phone numbers of most of my friends and my family.

At that time, when you got stumped with a question, you'd have to sit and try to connect your memory points to come up with the answer. You might have to phone a friend to see if they could remember. A lot of what we considered to be academic success was just around holding on to tidbits of information.

Then, the internet changed all that. Instead of marveling at the vastness of human knowledge, we slowly became accustomed to having answers at our fingertips. A simple Google search could provide us with information on virtually any topic, instantly. That meant the smartest person in the room was no longer the one who'd memorized all the information, it was the person with the fastest internet connection. Our focus shifted and intelligence moved from memorizing information to being better at finding information.

While the age of Google has undoubtedly made life easier, it's also diminished our sense of wonder and exploration. We stopped trying to figure it out in our heads, and instead tried to think of our Google search query. We stopped remembering phone numbers and instead worried about remembering our phone's passcode.

I mean really, how often do you find yourself thinking about a question for an extended period of time before grabbing your smartphone? How many phone numbers and addresses do you bother to remember with a supercharged computer in your pocket?

In truth, we've become so reliant on technology and the ease of access that it's provided, that we've forgotten how to wonder and how to question.

And now, we stand at another pivotal moment in human history. The rise of artificial intelligence, particularly large language models, isn't just changing how we find answers - it's fundamentally transforming the very nature of questioning itself.

We've moved from the industrial age, where knowledge was power, to the information age, where access to knowledge was power, and now we're entering the intelligence age, where the ability to ask the right questions is becoming the true measure of power and capability.

The AI tools at our disposal today aren't just search engines - they're conversation partners, analysis tools, and creative collaborators. But here's the fascinating part: their usefulness is directly proportional to the quality of our questions. A poorly-framed question

19

might get you a technically correct but practically useless response. A well-crafted question, on the other hand, can unlock insights that might have taken hours or even days to discover through traditional means.

This is why the next frontier in education isn't about teaching students how to find information - that's become trivial. It's not even about teaching them how to use AI tools - though that's certainly important. The next frontier is about developing what I call "question literacy": the ability to formulate, refine, and evolve questions that lead to meaningful discovery and understanding.

We have to learn to ask better questions.

When you're using ChatGPT or any other AI tool, the difference between getting a superficial response and a profound insight often comes down to how you frame your question. It's not just about asking "What is..." anymore. It's about asking "How might we..." or "What if..." or "In what ways could..."

This is where the PEACE Framework we'll be covering in chapter 4 becomes particularly relevant. Teaching with a question-first approach allows us to help students develop both the critical-thinking skills and technological fluency they need to navigate an AI-powered world. PEACE, which stands for **P**rovoke, **E**nquire, **A**nalyze, **C**reate, and **E**ngage, offers a structured yet adaptable framework for inquiry-based learning that embraces AI's role without letting it dominate. We'll unpack each component later. In the meantime, think of PEACE as a roadmap for transforming student curiosity into meaningful learning experiences, with AI as a tool and not a crutch.

So, what if the success of a lesson wasn't measured by what students could answer, but by what they chose to ask? How could our classrooms change if we valued curiosity as much as correctness?

After all, we're not just preparing them for a world where AI exists; we're preparing them to be architects of conversations with AI that lead to genuine discovery and innovation.

This shift brings us full circle from my childhood memories of Jeopardy. Today's champions aren't those who can store and recall vast amounts of information... They're the ones who can craft questions that unlock new possibilities, challenge assumptions, and push the boundaries of what we think is possible.

And that's exactly what makes this moment in education so exciting. As educators, our role has evolved from being providers of answers to being cultivators of curiosity and architects of inquiry. We need to help our students develop the skills to ask questions that challenge, inspire, and lead to innovation.

We've officially moved from "remember it" to "Google it"; from "Google it" to "ask it."

Before we dig into the PEACE Framework, we need to lay a few critical foundations. First, we'll explore the shift from memorization to mastery, showing that questioning is the true gateway to deep understanding. In the chapters that follow, we'll explore the fast-changing world of artificial intelligence in education by separating hype from reality and tackling the very real concerns that keep teachers up at night. We'll look at the difference between using AI as a simple automation tool (barely scratching the surface of its potential) and leveraging it for creative and analytical purposes that can transform learning. We'll confront the challenges schools face in integrating AI, from practical issues like access and equity to deeper concerns about academic integrity and the shifting nature of assessment.

After all, in a world where AI can instantly produce a five-paragraph essay, what becomes more valuable; the ability to write one or the ability to ask the kinds of questions that make that essay brilliant? How do we prepare students not just to work with AI, but to think beyond it?

Inquiry-based learning isn't new. It's a time-tested approach that has guided us for decades, but just as our students are facing a world fundamentally different from the one we grew up in, our approach to inquiry needs a thoughtful update. The principles of inquiry remain sound, but the practice must evolve to meet the needs of students who will spend their lives collaborating with AI.

Think of these next chapters as setting the stage for a new kind of classroom where technology and tradition don't compete but complement each other, where AI isn't just another tool but a catalyst for deeper learning, and where questions aren't just the starting point of inquiry but the driving force of innovation.

The Shift from Memorization to Mastery

Remember those old multiplication tables hanging on classroom walls? For generations, they were the cornerstone of what we thought learning was all about - memorizing facts until they became second nature. But how many of us still rely on those memorized tables to get to an answer? More likely than not, we turn to tech tools to get answers faster and leave the mental calculations behind. Most importantly, we understand the mechanics behind the math, which was the key lesson to take away, even if we defer to the tech tool.

This distinction between knowing facts and understanding concepts lies at the heart of education's great transformation. It's not just about the tools changing (though they certainly have); it's about our entire approach to what constitutes learning and mastery.

Think about learning to cook. You could memorize recipes word for word, following each step with military precision. Or you could understand the principles of cooking - how heat affects different ingredients, why certain flavors complement each other, how different techniques change textures. The first approach makes you dependent on recipes; the second makes you a chef.

This is where our educational paradigm is heading. With AI handling the role of our collective memory bank, we're free to focus on developing deeper understanding and practical application. Mastery isn't about regurgitating information anymore, it's about:

- Understanding underlying principles and patterns
- Recognizing how concepts connect across different domains
- Applying knowledge creatively to solve new problems
- Adapting strategies based on context
- Teaching others by articulating complex ideas simply

The beauty of this shift is that it aligns perfectly with how our brains naturally learn. We're not designed to be encyclopedias; we're designed to be problem solvers, pattern recognizers, and meaning makers. When we focus on mastery rather than memorization, we're working with our natural cognitive strengths rather than against them.

Consider how a master chess player thinks compared to a novice. The novice might try to memorize specific move sequences, while the master recognizes patterns and principles that apply across countless situations. This is the difference between surface-level learning and true mastery - and it's the direction education needs to take in the AI era.

But here's the fascinating part: Mastery doesn't make memorization obsolete; it just puts it in its proper place. When you truly understand something, the relevant facts tend to stick naturally because they're connected to meaningful concepts and practical applications. It's like how you never forget the rules of your favorite game - they're part of a larger system that makes sense to you.

The challenge for educators isn't to abandon traditional teaching methods entirely, but to reimagine them as a pathway to deeper understanding. When memorization becomes a tool for mastery rather than an end in itself, it takes on new meaning and purpose.

This shift demands more from both teachers and students. It requires:

- More emphasis on process than product

- Greater comfort with ambiguity and exploration
- Deeper engagement with "why" questions
- More authentic, real-world applications
- Increased focus on metacognition and learning strategies

The goal isn't to know all the answers... It's to know how to think about the questions. In a world where AI can instantly provide facts and figures, our unique value lies in our ability to understand deeply, think critically, and apply knowledge creatively.

And perhaps that's the most exciting part of this transformation: we're finally free to focus on what makes us uniquely human; our capacity for creativity, wisdom, and understanding. Memorization served its purpose in an age of information scarcity. But in an age of information abundance, mastery is the new currency of learning.

The Limitations of Unrefined Questioning

For today's kids, a world without Google, the internet, or Wi-Fi is unthinkable. They've never known life without it. But for those of us who lived through the rise of these technologies, we remember how the internet revolutionized the way we got information. Suddenly, with just a click and a question, the world's knowledge was at our fingertips. But learning how to ask the right questions... that was the challenge.

This meant to get the results you actually needed, you had to master search tactics, like quotation marks, plus signs, filtering by images, dates, and sources. You had to query like a boss.

And we saw this play out in the classroom. A student might type something like "what's the best job for me?" into Google and be met with a million results. If they were really motivated, they'd click on a few links, skim the content, and either get frustrated with the 'research process' or drown in a sea of "this isn't what I meant." Web pages responded by optimizing content generic enough to appeal to a wide audience, but just specific enough to seem helpful. For students, it became a game of asking, checking, and then checking again. The truth, even with all that power, a perfectly-phrased search could still return millions of irrelevant results, making it hard to zero in on the right information.

When voice-activated devices like Alexa and Siri arrived, it became even more obvious how crucial good questioning skills were. These tools had the potential to revolutionize how we access information and accomplish tasks, but their usefulness depends on how well we phrase our questions. A poorly worded command? That'll get you irrelevant (or downright bizarre) responses. To make things trickier, Alexa and Siri don't give you a list of options; they serve up what their algorithms think is the best answer. So instead of a million results to sift through, you get one answer... sometimes right, sometimes wrong.

Now, the gap between good and bad questions is widening fast, thanks to the rise of large language models (LLMs). Tools like ChatGPT are highlighting how crucial effective questioning really is. These models can produce human-like responses, but they're only as good as the directions they're fed. It's a classic case of "garbage in, garbage out."

Early on, many people who experimented with LLMs wrote them off as party tricks. They prompted (which is the process of asking or requesting a generative AI tool to perform a task), but the responses were often way off or sounded 'botty'. But here's the thing: It wasn't the tech that was faulty. The error was in how people questioned. The prompt mattered... a lot! A poorly-constructed prompt generated inaccurate, irrelevant, or even downright harmful responses.

Before we get too deep into the limitations of unrefined questioning, let's move this idea back to the classroom by playing a quick game. Think about the last time you asked your students to generate questions about a topic. What did you hear? I'm willing to bet it went something like this:

"What is..."

"When did..."

"Who made..."

Sound familiar? These aren't bad questions, per se. They're just... well, let's call them "Google-able." They're the fast food of questioning - quick, easy, and not particularly nutritious for the mind.

Here's the thing... When we first introduce AI tools like ChatGPT to our students, they tend to approach them the same way they approach Google. They ask these surface-level questions and get surface-level answers. Then, we wonder why they're not getting deeper insights or developing critical-thinking skills.

It's like handing someone the keys to a Ferrari and watching them drive it to the mailbox and back.

The problem isn't with the questions themselves; it's with our acceptance of them as good enough. We've gotten so comfortable with quick answers that we've forgotten how to dive deep. When a student asks, "What caused World War I," they'll get an answer, sure. But what about asking, "How might the lead-up to World War I parallel modern international tensions?" or "What alternative diplomatic solutions might have prevented the escalation of conflicts in 1914, and how could those lessons apply today?"

See the difference? The first question gets you facts. The second and third questions spark discussions, encourage critical thinking, and might even lead to insights that your AI companion hasn't considered before. On top of that, the second and third questions also require a student to learn about and understand the first question as well. In other words, they aren't missing that key factual information but processing it as part of a larger inquiry.

But here's where it gets really interesting (and a little uncomfortable): Many of us teachers aren't great at asking refined questions either. Don't feel bad; it's not entirely our fault. We came up through an educational system that rewarded having the right answer more than asking the right question. We learned to be answer-focused rather than inquiry-driven.

Let me share a personal teaching moment that still makes me cringe. I once asked my students to use ChatGPT to "explain photosynthesis." What I got back were twenty-six nearly identical responses that could have come straight from a textbook. What I should have asked was something like, "How might understanding photosynthesis help us design more sustainable cities?" or "What would happen to Earth's ecosystems if photosynthesis was 20% more efficient?"

These unrefined questions limit us in several critical ways:

- They encourage passive rather than active learning
- They focus on reproduction rather than application of knowledge
- They miss opportunities for cross-disciplinary connections
- They fail to engage students' natural curiosity and creativity

- They don't prepare students for real-world problem solving

Perhaps most importantly, unrefined questioning reinforces what I call the "vending machine approach" to AI: Insert question, receive answer, move on. It treats AI as a fact dispenser rather than a collaborative thinking partner.

The real power of AI in education isn't in its ability to provide answers. It's in its capacity to engage in an iterative dialogue that builds understanding, but this only happens when we move beyond surface-level questioning.

Think of it this way: If questions were tools, many of us are still using a hammer for everything when we have an entire workshop at our disposal. We need to learn to use the full range of tools, and more importantly, we need to teach our students to do the same.

In chapter four, we'll dive into the PEACE Framework, which will give you practical strategies for developing these more sophisticated questioning skills. But for now, I want you to start noticing the questions you ask and the questions your students ask.

Are they truly pushing thinking forward, or are they just Googling with extra steps?

In a world where AI can instantly answer most factual questions, the real value lies in asking questions that machines haven't even thought to ask yet. That's where human creativity and critical thinking truly shine.

Questioning as a Gateway to AI Proficiency

So, how do we help students navigate the growing complexities of our digital world? The answer is simple: by sharpening their questioning skills. And no, it's not to be better Googlers. It's about staying ahead of the AI revolution and ensuring our students are prepared to succeed in tomorrow's workforce.

Honestly, when was the last time you saw a job posting that didn't mention "problem solving" or "critical thinking" as required skills? These aren't just corporate buzzwords anymore. They're survival skills in a world where routine tasks are increasingly automated. The factory workers of tomorrow won't be assembling widgets; they'll be troubleshooting the AI systems that assemble the widgets. The office workers of tomorrow won't be processing data; they'll be asking the right questions to interpret what the AI-processed data actually means.

These skills are grounded in three essential pillars. First, there's critical thinking. Students need to analyze, synthesize, and draw conclusions from the flood of information available to them. It's not enough to ask AI "What's the answer?" Students need to ask "Why is this the answer?" and "How can I verify this?" and "What might be missing from this analysis?"

Next, there's problem solving. Students must be able to break down complex challenges into manageable, bite-sized steps. This isn't just about finding solutions; it's about defining the problem correctly in the first place. After all, AI is only as good as the problem you present to it. I've seen countless students struggle with AI not because they can't use the technology, but because they haven't properly identified what they're trying to solve.

Finally, there's prompt engineering. That's right, learning how to craft effective directions for AI interactions is becoming a critical skill. And as AI becomes an even bigger part of everyday life, this skill will only grow in importance. Think of it as learning a new language - the language of AI collaboration. Just like you wouldn't walk into a French restaurant and order in English (well, you could, but good luck getting exactly what you want), you can't approach AI with vague or poorly-structured prompts and expect optimal results.

You know that student who always asks the perfect question that makes the whole class go "Ohh..."? That's what we're aiming for with AI interactions. We want students who can craft prompts that make AI tools go "Ohh..." (metaphorically speaking, of course - let's not get too carried away with AI personification).

But here's the real kicker: These skills aren't just about academic success or even career readiness. They're about empowerment. In a world increasingly mediated by AI, the people who can ask the right questions will be the ones shaping the future. They'll be the ones turning AI from a tool into a partner, from a calculator into a collaborator.

These are the skills students need to thrive in the AI-driven world. It's up to us to help guide them there. And no, the irony isn't lost on me that we're using age-old questioning techniques to master cutting-edge technology. But maybe that's the point… Some skills are timeless, even if the tools we use to apply them keep evolving.

The challenge now isn't just teaching students to ask better questions; it's teaching them to ask better questions in a way that both humans and AI can understand and build upon. It's about bridging the gap between human curiosity and machine capability. And trust me, that's a lot more exciting than teaching them to fill out another worksheet - AI-generated or otherwise.

Coming Up

This is more than just equipping students with questioning skills. The real challenge is transforming how we teach, ensuring that instruction isn't just delivering content, but building the capacity for students to actively engage, question, and create with AI. After all, the goal isn't to use AI as a crutch that leads to passive learning, but rather as a tool that encourages deeper thinking, creativity, and problem solving.

Are you ready? In this book, we'll get into practical strategies to make this happen - how you can adjust your instruction to build questioning skills and create activities that use AI to build capacity, not scaffold passivity. If you're impatiently wondering how exactly to start adjusting instruction to improve questioning skills, or what types of AI activities can boost students' capacity instead of promoting passivity, then jump to Chapter 4 to get started with the PEACE Framework. That's where I break down exactly how to implement these strategies in your classroom. In the next few sections, we start by laying some foundation and addressing some AI concerns. Don't worry... by the time we're done, you'll have a toolbox full of practical ideas to use immediately, more confidence about AI as a learning tool, and the step-by-step to implementation.

But here's a question for you: What happens when students don't just ask better questions, but use AI to answer questions that haven't even been asked yet? Stick around, because we're about to explore just that.

Chapter 2

Understanding AI: Not Just Another Tool

I still remember the thrill of getting a few boxes of chalk as a welcome gift at my very first teaching job - seriously, I was excited! Did I mention the box of pastels? There was something about holding those chalky little sticks that made me feel like a "real teacher." And back then, using a computer in the classroom? That was a novelty - a little splash of tech in a sea of overhead projectors and textbooks.

Fast forward to today, and it's almost incomprehensible to think of a classroom without technology. The evolution has been fast and furious, transforming everything from how we communicate to how we learn. Now AI is leading the charge in this ongoing journey, changing the way we think about teaching.

39

When I think back to how much of my weekends and late nights were wrapped up in writing lesson plans, sending emails, and grading papers, it's a wonder I don't look a million years old! Since OpenAI dropped their ChatGPT large language model (LLM) bomb on us, I've readjusted everything I do in my professional life, and I mean everything. As an early adopter and ed tech lover, I cannot imagine teaching without it, just like that box of pastel chalk.

But, let's pause for a second and really take this in: With the rapid rise of AI, especially tools like LLMs, we can't hold steadfast to the idea of students passively absorbing information from textbooks or even from their teachers. Instead, look forward to virtual assistants by your students' side, ready to answer their questions, give personalized feedback, or even suggest new learning paths tailored to their needs. AI is catapulting from a handy tool for automating tasks into a creative thought partner in the learning process.

But here's the thing: With all the excitement surrounding AI, there's also a fair share of confusion. Myths and misconceptions seem to swirl around it like chalk dust in the air (trust me, I know how that stuff lingers). So, what is AI really? And how do we separate reality from the hype?

Let's start with what AI isn't. It's not a magic wand that will suddenly make all our teaching problems disappear. It's not going to replace teachers (despite what some sensationalist headlines might suggest). And it's definitely not going to turn our students into passive consumers of machine-generated knowledge.

Think of AI more like having a really knowledgeable teaching assistant who never sleeps, never gets tired, and is always ready to help - but who also sometimes gets things hilariously wrong and needs constant supervision. You wouldn't let a TA run your classroom without guidance, right? The same goes for AI.

What makes AI fundamentally different from previous educational technologies is its ability to adapt and respond. Your overhead projector never learned from its mistakes (though mine certainly made plenty). Your SmartBoard never suggested a different approach when students struggled with a concept. But AI can do both - and more.

Remember that student who always asked "why" about everything? The one who made you think deeply about concepts you thought you knew inside and out? AI is kind of like having that student in your pocket, constantly pushing you to clarify your thinking and consider new perspectives. Except this student comes with an off switch (wouldn't that have been nice?).

The real power of AI in education isn't in its ability to replace traditional teaching methods - it's in its capacity to enhance them. It's about taking the best of what we know about teaching and learning and supercharging it with technology that can adapt to individual needs, provide instant feedback, and create opportunities for deeper learning.

But here's where many educators get stuck - they try to fit AI into their existing teaching framework, like trying to jam a round peg into a square hole. They use it to generate worksheets faster or grade multiple-choice tests more quickly. And while there's nothing inherently wrong with that (efficiency is great!), it's like using a smartphone just to make phone calls - you're missing out on so much potential.

To really understand AI's role in education, we need to shift our thinking from "tool" to "environment". Just as we create physical learning environments in our classrooms, we need to think about creating AI-enhanced learning environments that promote inquiry, creativity, and critical thinking.

In the next sections, we'll explore exactly how to do that, looking at practical ways to integrate AI into your teaching practice that go well beyond the basics. But for now, just remember: That box of chalk didn't make you a teacher - it was just one tool in your arsenal. AI is the same way, just with a lot more possibilities (and thankfully, a lot less dust).

***

The Evolution of AI in Education

Let's take a quick stroll down memory lane. Artificial Intelligence (AI) has been around in education longer than most of us probably realize. It all started back in the 1960s when some bright minds at the University of Illinois decided to create Programmed Logic for Automatic Teaching Operations (PLATO). Students interacting with a computer to learn through text and graphics? Whoa! Sounds like historical science fiction, right? Well, it was a big deal back then and laid the groundwork for what we now consider modern educational technology.

Fast forward a couple of decades to the 1970s and 1980s, when Intelligent Tutoring Systems (ITS) began popping up. These systems were like the early versions of personalized learning. They used AI to tailor lessons based on how well students were doing. If you struggled with math, the system would throw in some extra practice problems! Meanwhile, natural language processing was making strides, leading to language learning programs that allowed students to chat with their computers.

Despite these cool innovations from decades ago, AI in education didn't exactly take off. Why? Well, technology was expensive and not everyone had access to it. Plus, early AI relied heavily on rigid programming, making it much less flexible than more recent iterations.

As we moved into the 1990s and beyond, things started to change. Researchers began forming communities focused on how AI could really make a difference in classrooms. They emphasized using data to guide decisions and personalized learning pathways that actually worked for diverse learners.

Fast forward to today. AI is everywhere in education. Tools like ChatGPT, Magic School AI, SchoolAI, and Khan Academy's Khanmigo are changing how we interact with technology, teach, and learn. These platforms provide tailored instruction, helping students engage with content in ways that suit them best. Plus, they lighten the load for us educators by automating some of those pesky administrative tasks.

So, what's the takeaway? AI has come a long way from those early computer systems. And, just like so many technology tools that came before it (like calculators, word processors, interactive whiteboards, and learning management systems), teachers cannot drag their feet in adopting it in the classroom. Why? Well, this version of AI that we use today is the dumbest AI our students will ever work with. As revolutionary as it was when Open AI dropped ChatGPT in the fall of 2022, it's already archaic as far as modern AI is concerned. We have to reshape how we approach education for all learners. As K-12 educators, especially those of us focused on instructional technology and inclusion, we have an incredible opportunity to give these bridge learners (the ones who will wax on poetically about life before AI) a big advantage. We can harness AI's potential and make learning more accessible and engaging for our students.

***

Debunking Misconceptions

Remember when everyone thought the internet was just a fad? (Surprise! It wasn't.) Well, now we're seeing the same kind of confusion around AI. It's become the educational equivalent of avocado toast – everyone's talking about it, but not everyone knows what it's actually all about. Let's cut through the noise and tackle some of these myths head-on, because understanding what AI isn't is just as important as understanding what it is.

Myth 1: AI Can Think and Feel Like a Human

Let's get real for a second. AI has about as much genuine emotion as my coffee maker. Sure, it might tell you it's "happy to help" or "sorry for the confusion," but that's like my coffee maker beeping to tell me my coffee's ready - it's just following its programming. When you're chatting with ChatGPT, you're not having a heart-to-heart with HAL 9000; you're interacting with an incredibly sophisticated pattern-matching system. Think of it less like a sentient being and more like the world's most advanced autocomplete. Impressive? Absolutely.

Ready to replace your school counselor? Not even close.

Myth 2: AI Understands Content Like Humans Do

Here's another one: Many people believe that AI truly understands what it's saying. Spoiler alert - it doesn't. While AI can process language and generate responses that sound natural, it lacks the deep comprehension we humans possess. It doesn't "get" context or nuance the way we do; it just follows patterns and probabilities. When AI generates a response about Shakespeare, it's not contemplating the deeper meaning of "to be or not to be" - it's analyzing patterns in countless Shakespeare-related texts and constructing a response that statistically makes sense. So, when you ask it a question, remember that it's not having an epiphany; it's just crunching numbers.

Myth 3: AI Will Take Over All Jobs

Ah, the classic fear of the robot uprising. Will AI really steal all our jobs? Let's get real... We're way past debating whether AI will be part of our work lives - that ship has sailed, my friends. The question isn't if AI will replace jobs, it's how we'll adapt to working alongside it. Think about it like the introduction of computers in the workplace. Sure, some jobs disappeared (pour one out for the typewriter repair folks), but how many people today work without using a computer? The reality is that AI isn't here to steal our jobs... it's here to transform them. The most likely scenario isn't a robot taking your desk, but rather AI becoming your always-on digital colleague that takes over the repetitive tasks so you're free to focus on the uniquely human aspects of your work. Studies suggest that while AI will absolutely disrupt certain job categories, it's more likely to augment most positions rather than eliminate them entirely. The real question we should be asking isn't "Will AI take my job?" but rather "How can I learn to leverage AI to become better at my job?" Because in tomorrow's workplace, the divide won't be between humans and AI - it'll be between humans who know how to work with AI and those who don't.

Myth 4: AI Is Completely Unbiased

Remember that one teacher who swore they were "totally neutral" during a heated team meeting debate, while clearly rooting for their best friend's position? AI's kind of like that. Despite what the tech evangelists might tell you, AI isn't some perfectly objective digital oracle. It's more like a mirror reflecting all the messy human biases in its training data. Those thousands of internet articles, books, and documents it learned from? Yeah, they came with all our human baggage included. So, when your students are using AI tools, teach them to be like good journalists - verify, question, and always consider the source. Because if we're not careful about checking for bias, we might end up with the digital equivalent of that one textbook from 1952 that's still somehow in your classroom cabinet.

Myth 5: Once Built, AI Can Work Alone

Wouldn't it be nice if AI was like one of those set-it-and-forget-it slow cookers? Just turn it on and come back to perfect results? Unfortunately, AI is more like having a student teacher who needs constant mentoring. It requires ongoing guidance, updates, and quality checks. And here's a wild thought: Now that AI has been around long enough to start training on its own outputs, we're seeing something like a game of digital telephone. Each generation of AI learning from the last might be like that worksheet that's been photocopied so many times you can barely read it anymore. The lesson? AI needs human oversight just like your students need your guidance - consistently and thoughtfully.

Myth 6: AI Is Only for Tech-Savvy People

If you can figure out how to use the temperamental copy machine in the teacher's lounge (you know the one), you can figure out how to use AI. Seriously. Today's AI tools are about as complicated as ordering pizza online - and probably less frustrating than trying to get that copy machine to collate properly. You don't need to know coding or have a secret handshake with Silicon Valley insiders. If your students can navigate TikTok, they can navigate AI tools. The key isn't technical expertise; it's curiosity and willingness to experiment. Moreover, AI has slowly been integrated everywhere. Suggesting things to buy from your Amazon purchase history, playing music based on your listening patterns, and predicting what you're going to say next in a text message... we've all been using AI for a while without realizing it. Still, for those people who think AI is like programming a supercomputer, breathe. Start small, experiment, and build from there.

Myth 7: AI Can Replace Human Creativity

Finally, let's tackle the idea that AI can outshine human creativity. Here's a fun experiment: Ask AI to write a poem about your morning coffee, then ask your third-period class to do the same. Sure, the AI might give you something technically perfect with flawless meter and rhyme, but it'll be missing that special something - like that one student's poem about how their mom's coffee mug collection is taking over the kitchen. AI can remix existing patterns into something new, but it can't capture the genuine human experience that makes creativity truly powerful. The magic happens when we use AI to amplify human creativity, not replace it.

Myths Debunked!

Understanding these myths isn't just about separating fact from fiction - it's about seeing AI for what it really is: A powerful tool that, like any good teaching resource, works best when guided by human wisdom and experience. Now, shall we talk about how to actually put this understanding to work in your classroom?

***

AI for Automation vs. a Partner for Creative Learning

Remember when grading papers felt like an endless mountain of work? (Who am I kidding - it still does!) AI can now swoop in and handle those tedious tasks, freeing us up for the real magic - connecting with students. But here's the best part: AI isn't just your virtual teaching assistant who helps you tackle your piles of paperwork, it's also that creative colleague down the hall who always has fresh ideas and never minds if you pop in for a brainstorming session, even at 10 PM on a Sunday.

The question is, how do we make sure we're using it for the right reasons - both automation and creativity? Let's break it down.

AI as a Tool for Automation

Let's start with the obvious win: Automation! AI can absolutely be the best assistant ever... the one who doesn't need coffee breaks or complain about paperwork. Think of all those mind-numbing tasks that make you question your career choices - grading multiple-choice tests, writing detailed lesson plan outlines, or creating that perfect rubric for the fifteenth time this year.

These AI tools are like having the world's most efficient teaching assistant. They can grade those quizzes faster than you can say "professional development day," analyze student data without getting cross-eyed, and even help you draft those dreaded progress reports. I've always wished there were more of me to help me get my job done. Now, there is, minus the unhealthy need for caffeine.

When you use AI for automation, it's about clearing the clutter so you can focus on the magic. You know, those "aha!" moments that made you want to become a teacher in the first place.

If you are new to AI prompting and the idea of using LLMs to build better classroom resources but need help with differentiation and how to support streamlining your workload, then join me for an online course all about teacher hacks with AI at www.peaceframework.com/promptai

AI as a Partner for Creative Learning

Now, this is where it gets exciting! Imagine having a co-teacher who never runs out of ideas, never gets tired of brainstorming, and doesn't mind if you shoot down their suggestions. That's AI as a creative partner.

Instead of just asking AI to grade papers or write lesson plans (yawn), we can use it to spark creativity and innovation. It's like spit balling with a mentor who's read every education book ever written and somehow still has the energy to help you design that perfect project-based learning unit at midnight.

Want to differentiate learning for your class? AI can help create personalized paths that make your students feel like the content was made just for them (because, well, it was). Need to group students for a project? AI can play matchmaker better than your aunt at a family reunion, pairing students based on complementary skills and learning styles.

And here's a wild thought: What if we let AI help students discover their own learning superpowers? It can adapt to their pace, suggest resources that actually interest them (goodbye, one-size-fits-all worksheets!), and create interactive content that makes learning feel more like exploration than obligation. But in a way that's more accessible and responsive to diverse learning needs than any other ed tech tool that's come before it.

Classroom Vignette: Understanding AI is Not Just Another Tool

In Mr. Chen's high school chemistry lab, students cluster around their lab benches, faces animated as they debate their latest findings. The room is noisy and buzzes with the kind of energy that comes from active investigation.

The Setup

"I was drowning in lab reports," admits Mr. Chen, laughing as he remembers his initial skepticism about AI. "I kept thinking - there has to be a better way to help students understand the scientific method while keeping my sanity."

Idea in Action

Mr. Chen uses AI in two distinct ways. First, for efficiency: He created prompt templates for providing detailed feedback on lab reports, focusing on methodology and analytical thinking. The second way was much more exciting. He was able to light a fire of engagement by using AI to generate "mystery observations" for his students to investigate.

Sample Teacher Workflow

1. Before class, Mr. Chen inputs basic parameters about the day's topic (e.g., "Generate 3 unusual but scientifically accurate observations about chemical reactions in everyday life")

2. Students receive these observations and must:

   o Design experiments to test if they're true
   o Use AI to research potential explanations
   o Compare AI-generated hypotheses with their own
   o Document where AI helped and where it led them astray

Teacher Reflection

"Last week, one AI observation suggested that copper pennies turn green for the same reason leaves do. My students were immediately skeptical... They've developed this wonderful critical sense. They used AI to research both processes, designed experiments to test the claim, and ended up teaching me something new about oxidation. The real breakthrough wasn't just in their understanding of chemistry; it was in their ability to question and verify information, even from AI sources."

Finding the Balance

So how do we strike a balance between these two roles? How do we walk this tightrope between efficiency and creativity? AI is our Swiss Army knife... It has lots of tools, but you're the one deciding which one to use and when. You can't do much of anything using the wrong tool (like the screwdriver to cut paper). When it comes to AI, use it to handle the paperwork avalanche, sure, but also let it be your creative springboard for those "what if" moments that make teaching exciting.

Remember: AI might be smart, but it doesn't have your teacher instincts. It can't give a struggling student that encouraging smile or celebrate a breakthrough moment with a perfectly timed high-five. Your emotional intelligence, intuition, and ability to connect with students - that's your superpower, and no amount of artificial intelligence can replace that.

The future of education isn't about choosing between human teachers and AI - it's about combining the best of both worlds. Let AI handle the grunt work and help spark creativity, while you focus on what you do best - being that irreplaceable human presence in your students' learning journey.

So, what's next for your classroom? Maybe it's time to let AI take over your Sunday night grading sessions while you focus on designing that amazing project you've been dreaming about. After all, the future of education is whatever we decide to make it - we might as well make it awesome.

***

Challenges of Integrating AI in Schools

Let's hop in a time machine. Go back to a time when getting an overhead projector in your classroom felt like a technological revolution that would change everything! Fast forward to today, and we're grappling with AI tools that are supposed to change everything. *Again*. Except this time, it's not just about shining light on the screen; it's about shining a light on how we learn, teach, and assess. And if that feels like a big leap, well... it is. The truth is, integrating AI in schools isn't as simple as downloading an app or plugging in a new device. It comes with a set of challenges that feel less like "plug and play" and more like "trial by fire."

But let's face it, no one ever said revolutionizing education would be easy, right?

Common Fears: Cheating, Replacing Teachers, and Dependency

When we take huge leaps and work to reframe our approach to education, it's only natural to encounter some bumps along the way. While the benefits of AI are exciting, fears about cheating, replacing teachers, and student dependency can create some serious roadblocks. Let's unpack these concerns and see how we can tackle them.

Fear 1: Cheating

Ah, cheating. It's the teacher's perennial nemesis. Except now, it's turbocharged with the power of AI. I mean, imagine your students turning in essays that George Orwell himself couldn't tell apart as human or machine. That's the fear, isn't it? That students will use AI to shortcut their way to an "A" without really learning anything.

But here's the thing: Cheating existed long before AI, and it's not going anywhere. Does that mean we just throw our hands up in defeat? Not quite. But we certainly can't stick with the same old assessments that are easy to game. We've got to get a little creative because our students need AI-ready skills to compete in future job markets. We can't keep AI locked in the supply closet for fear of the cheaters.

What do we do instead? Let's rethink how we assess learning. If we're still relying on easily automated tasks, like basic multiple-choice tests or generic essays, we're setting ourselves (and our students) up for failure. Instead, we need to lean into assessments that require creativity and critical thinking. Want to know if your students understand the material? Have them create something - whether that's a presentation, a podcast, or even a digital portfolio that AI can't replicate. Try in-class assessments, where students can collaborate and discuss their thinking processes in real time. By shifting from "regurgitate this" to "create that," we make it nearly impossible for AI to do all the heavy lifting.

Fear 2: Replacing Teachers

Next up is the fear that AI will take over our jobs. I mean, who wants to be replaced by a robot?

Now, here's a fun one. Will AI swoop in like some tech-savvy villain and replace us teachers? Yeah, no. Remember when calculators were going to make math teachers obsolete? Or when spell-check was supposed to put English teachers out of work? How about when Google was going to make librarians unnecessary. How'd that all work out?

Here's the thing: Teaching is more than just report card comments and lectures. Yes, AI is good at grading quizzes and spitting out lesson plans faster than you can say "rubric," but can it give a pep talk to a struggling student or offer a high-five after an "aha" moment? Didn't think so.

AI isn't here to steal the spotlight. It's meant to support us, and it can only accomplish what we tell it to do. Someone has to be driving the bus here, and it's you! At the end of the day, no algorithm can ignite a student's passion for learning quite like a human teacher.

What do we do instead? Let's reframe our thinking: AI is like a teacher's aide that never takes a sick day. It can handle those tedious tasks so we can focus on building relationships with our students and creating a positive learning environment for all. As facilitators of the learning process, teachers are irreplaceable. With that said, you have to be ready to immerse yourself in learning about AI and integrating new teaching methodologies into your classroom. No rolling in the TV cart and tape player to show off your teaching chops. This is a moment of fundamental change in the landscape of AI. If you want to grow as a teacher, you'll have to shift with it. The fact that you're reading this book says a lot, but it's only the first step. Take it to the next level by applying these strategies in your classroom and updating your instructional pedagogy. Your students are dependent on you to teach them the skills they need for the tasks of tomorrow.

Fear 3: Dependency

Finally, we fear dependency, and this is where things get tricky. If we let students lean too heavily on AI, what happens to their critical-thinking skills? Will they even know how to solve problems without asking ChatGPT for a lifeline? It's a valid concern, right? The fear is that we're turning kids into walking extensions of their devices.

What do we do instead? Let's flip the script for a second. What if, instead of fearing dependency, we taught them how to *partner* with AI? Kind of like when we first taught students how to use Google. Remember that? We didn't ban it because it gave answers too easily. Just like we teach students to navigate the internet with a critical eye, we can guide them to use AI thoughtfully. Encourage them to ask deeper questions - not just "What's the answer?" but "Why is this the answer?" and "How could I use this information differently?"

The PEACE Framework, which we'll cover in chapter 4, can help you guide students to use AI thoughtfully, not as a crutch, but as a springboard for deeper discovery. And that's a skill worth teaching. After all, they're not going to stop using AI anytime soon - so why not show them how to do it right?

While fears about cheating, teacher replacement, and student dependency are common in conversations about integrating AI in schools, they don't have to hold us back. By addressing these concerns head-on - with authentic assessments, emphasizing our irreplaceable role as educators, and promoting balanced use of technology - we can create an environment where AI enhances education instead of detracting from it. So, let's keep these challenges in mind as we continue exploring how to make AI work for us in our classrooms.

Ethical Considerations

Is it just me, or does letting AI handle all that data feel a bit like handing over your deepest secrets to a stranger at the bus stop? I mean, who wouldn't feel a little uneasy about where all that personal information is going? If you've ever found yourself wondering, "What happens to all this data once I hit enter?" -

you're not alone. But here's the deal: Just like we teach our students to be responsible digital citizens, we have to take on the role of privacy protectors in the classroom. The challenge? Balancing innovation with protection. So, let's break it down into three big concerns - privacy, bias, and data security.

Privacy

Ah, privacy. The very word brings up images of someone peeking over your shoulder and looking at your phone. Excuse me! With AI collecting data on everything from student performance to learning preferences, it's easy to imagine a Black Mirror scenario where all this info is swirling in some giant data cloud, just waiting to fall into the wrong hands. So, how do we keep our students from feeling like they're under some AI-powered microscope?

Transparency, transparency, transparency. Make it crystal clear to students, parents, and even your colleagues about what data is being collected and how it's being used. Establish policies that prioritize privacy. Think FERPA (Family Educational Rights and Privacy Act) compliance, student initials instead of full names, and only sharing the bare minimum of necessary data. In short, don't tell AI everything. Keep it on a need-to-know basis. After all, you wouldn't tell someone your home address just because they asked, right? Same goes for student data.

Bias

Let's get real - AI is only as "unbiased" as the data it's trained on. And if that data reflects human biases (spoiler alert: it does), then AI is going to reflect those same biases right back at us. It's not just about algorithms gone awry, either. AI can reinforce stereotypes, gender roles, and even disability bias with something as seemingly innocent as a phrase like "against all odds," which subtly implies that overcoming a disability is some kind of heroic exception rather than a reflection of diverse human experiences. The last thing we want is for AI to amplify the very inequalities we're working to dismantle, especially for marginalized students or those with diverse learning needs.

So, what's the solution? Be proactive, not reactive. Take a close look at the AI tools you're using and see if they prioritize fairness and inclusivity. When you're using AI-generated content, go ahead and weave fairness into your prompts. Be specific in guiding AI toward inclusive language, and watch for the subtle biases that creep in. And don't be shy about bringing your students into the conversation. Teaching them to recognize bias in AI not only makes them more critical consumers of information but also prepares them for a world where tech isn't always as neutral as we'd like to think. If they can spot bias in their TikTok feeds, they can definitely handle it in an AI tool.

Data Security

Ever wondered what happens to all those questions you ask your favorite chatbot? Well, it's like feeding a supercomputer a steady diet of your deepest thoughts and biggest dreams. And while the end result is awesome, it also raises a big question: Is our data safe? With so much valuable information floating around the digital universe, we need to make sure we're locking down our data fortresses. After all, even the most groundbreaking innovations aren't worth it if the price is our privacy.

When it comes to our students' information, we need to be extra careful. Feeding their personal details or academic records into AI tools might seem harmless, but it's like leaving their files open on the teacher's lounge table - it's not just a breach of trust, it's the kind of mistake that could have serious consequences, both ethical and legal.

So, what do we do instead? Keep it simple and secure. First up, let's play by the rules. Those district guidelines about data privacy aren't just suggestions (even if we sometimes treat the dress code that way). Many districts now have their own AI contracts with providers who promise not to collect or store personal information - use these approved tools when you can. When you're using AI tools on your own, think protective and professional. Instead of typing "Jimmy in third period is struggling with fractions," try "I have a student who's having trouble with..." Small change, big difference! The key is avoiding personally identifiable information - no student names, no specific details about their lives. Keep it anonymous, keep it safe.

And please, for the love of all things educational, update those passwords! Adding an exclamation point to "password123" isn't fooling anyone. I know, I know - between scarfing down your slightly-too-cold lunch leftovers and prepping for afternoon classes, who has time for security updates? But trust me, dealing with a data breach would eat up *way* more time than taking five minutes to set up two-factor authentication or updating your browser. Protecting our student's well-being doesn't stop at the classroom door. It extends into every digital space we use, which means securing students' digital information is just as important as making sure they're safe in our classrooms.

Conclusion

When we talk about integrating AI into education, it's easy to get caught up in the shiny possibilities - personalized learning, innovative assessments, less paperwork (yes, please!). But we can't forget that with great tech comes great responsibility. When it comes to challenges of integrating AI, we can overcome our fears and usher in this new era. Whether it's ensuring privacy by being transparent about what data we're collecting, tackling bias by critically evaluating the tools we use, or prioritizing data security to protect against breaches, it's on us to build a safe, inclusive, and ethical learning environment.

Because at the end of the day, AI is just a tool. It's up to us to make sure we're using it in ways that support our students, keep them safe, and empower them to navigate an increasingly digital world. And if we do that right, the possibilities are endless.

***

Navigating the AI Landscape

As we've unpacked in this chapter, AI isn't just another gadget we can toss into the classroom mix - it's a game changer, a force that's already reshaping how we teach and learn. But with great transformation comes a few speed bumps along the way. We've seen how AI's promise comes with its own set of ethical dilemmas, privacy concerns, and the ever-present question of fairness.

So, how do we move forward without falling into the trap of "tech for tech's sake"? It's about finding that sweet spot where we lean into AI's power and potential without losing sight of what makes learning in our classrooms feel human. We need to be clear-eyed about AI's strengths and its limitations while also setting thoughtful parameters that balance the brilliance of automation with the irreplaceable magic of human interaction.

The road ahead? Well, it's filled with questions - and that's a good thing. How do we ensure that asking the right questions remains at the heart of learning, even in an AI-powered classroom? Will our students' ability to think creatively and problem solve become even more valuable, or will they lean too heavily on technology for answers? And perhaps the biggest question of all: How do we make sure our students are ready not just to live in this AI-driven world, but to lead in it?

Chapter 3

Inquiry-Based Learning

Remember those old-school study sessions where you and your friends spent hours making flashcards, desperately cramming dates, formulas, and facts into your brains, hoping they'd stick for the big test? Luckily, those days are fading fast. We've already established that memorizing Googleable facts is out and question-driven learning is in. Before we dive into the PEACE Framework as a way to AI-power traditional inquiry-based learning (IBL), let's review a few key principles from the unplugged version first.

Traditional IBL Model Explained

When we talk about a traditional inquiry-based model, we're working with a structured approach that guides students through the process of exploration and discovery, often with a scientific lens. It's a time-tested method for helping students learn how to investigate, problem solve, and draw conclusions on their own. Although this model can be summarized in five main phases, it actually includes eight distinct parts, each with a clear purpose and flow to keep students moving forward in their inquiry. Here's how it all breaks down:

Step 1 of Traditional IBL
Orientation

It starts with orientation, which is essentially an invitation to explore. In this phase, *teachers introduce students to the topic*, problem, or phenomenon that will anchor their inquiry. It's all about setting the stage for the learning ahead. Teachers might use demonstrations, intriguing videos, or experiments to get students wanting to dig deeper. It's also an opportunity to share critical information essential to the target learning objectives. The ultimate goal of orientation is to set the stage for student-generated questions that will guide students' learning throughout the inquiry process.

Step 2 of Traditional IBL

Conceptualization through Questioning and Hypothesis Generation

Once oriented, it's time for conceptualization - a phase with two important parts. First, students begin *formulating questions* on their own. This isn't about asking the teacher for answers; it's about learning to ask questions that open up new avenues for discovery. Teachers play a key role here, helping students refine and focus their questions to make them investigable. The second part of conceptualization is *hypothesis generation*, where students take a stab at predictions or explanations based on what they know so far. With teacher support, they work to identify variables and consider relationships between different ideas. It's here that the groundwork is laid for meaningful investigation, turning broad curiosity into focused inquiry.

Step 3 of Traditional IBL

Investigation through Exploration, Experimentation, and Data interpretation

Now, we're getting to the heart of the inquiry process - investigation. In this phase, students do the active work of testing their hypotheses through exploration, experimentation, and data interpretation. It's a three-part stage where students roll up their sleeves and get hands-on. First, they *plan and conduct experiments or research* activities to test their hypothesis. *Data collection* follows, with students gathering observations, measurements, survey results, or whatever fits their investigative approach. Finally, they *analyze and interpret that data*, often with the help of technology to streamline their efforts. And if things don't go quite as planned - they're encouraged to revise their methods, embracing trial and error as part of the journey. This phase builds resilience and flexibility, showing students that roadblocks are just a chance to try something new.

Step 4 of Traditional IBL

Conclusions

In the conclusions phase, students bring everything together. Here, they evaluate their findings, *drawing conclusions* that either support or refute their original hypothesis. It's not about "getting it right;" it's about understanding what the data says. This stage encourages them to identify patterns, uncover relationships, and consider alternative explanations when things don't add up. By synthesizing their findings and evaluating the evidence, students learn that meaningful inquiry often leads to more questions than answers - a true sign of discovery.

Step 5 of Traditional IBL
Discussion through Communication and Reflection

The final stage of traditional inquiry-based learning centers on *sharing findings* and reflecting on the process. Students are allowed choices so they can play into their strengths and get creative by using presentations, reports, posters, or digital projects to express learning meaningfully. After presenting, peers and teachers provide feedback, ask questions, and may challenge findings.

Reflection is also a key element of this fifth step. Students review their inquiry process, considering what worked, what was challenging, and how they could improve. This reflection deepens learning as they learn from successes and recognize the value of overcoming obstacles. Ultimately, this stage is about stepping back from the hands-on exploration, engaging in a meaningful dialogue about what was discovered, and connecting new knowledge to a broader understanding. It can also serve as a new jumping off point for further inquiry, making it cyclical. This mirrors the scientific process, showing students that learning is an ongoing journey where answers often lead to more questions, and curiosity is never fully satisfied.

The Benefits of Inquiry-Based Learning

Setting up inquiry-based learning sounds easy enough, right? You might even be doing elements of this model already without realizing it. And honestly, there are plenty of reasons why teachers are drawn to this approach. It just works.

1. IBL Cranks Up Engagement

Keeping students engaged can feel like a losing battle, especially when you're up against the latest viral TikTok video. Why is social media so addictive? AI algorithms are curating suggestions based on search history and watch time, meaning it's personalized! It caters to the interests of each student, pulling them down rabbit holes they actually care about. And that's exactly where inquiry-based learning shines. When you let students explore questions and topics that matter to them, they get invested. Suddenly, school isn't just about filling in worksheets; it's about exploring what makes them curious.

This kind of authentic engagement - it's a game-changer. Not only does it lead to more participation in class, but it also boosts attendance, improves the quality of student work, and yes, even ramps up those all-important learning outcomes. When students are hooked on their own learning, they show up and show out - and not in the disruptive way. Actually, this kind of interest-based learning does wonders for minimizing classroom behaviors. Here's why: When students are deeply engaged in something that resonates with them, they're less likely to act out. Instead of becoming distractions, they're too busy digging into their projects to bother with off-task behavior.

And let's not ignore the elephant in the room (yes, the one shooting spitballs in the corner). Behavior management has been harder than ever since that never-ending "spring break" of 2020 threw emotional regulation and executive functioning out the window. Inquiry-based learning pulls double duty: It keeps students engaged while helping them rebuild self-management skills. When they're deep in meaningful work, they're not just learning about their topics; they're also practicing focus, persistence, and collaboration, all of which help cut down on classroom disruptions.

2. IBL Builds Critical Thinkers

Let's be real - critical thinking isn't just a nice-to-have skill; it's essential. In a world where information is endless, AI content is everywhere, and the line between fact and fiction seems blurrier by the day, our students need more than just surface-level knowledge. They need the skills to sift, sort, and scrutinize information. Whether they're parsing through AI-generated content or navigating the overwhelming flood of information that comes with every Google search, the ability to evaluate, analyze, and draw informed conclusions is a non-negotiable. And this, right here, is the heart of inquiry-based learning.

Right now, in many classrooms, we're still stuck in a cycle of memorization or, at best, "Google it." But here's the catch - our students might be digital natives, but they're often far from digital experts. They're great at typing a question into a search bar but not so great at querying effectively. Most don't use thoughtful search terms to get the best results and, in reality, they're likely to click on the first link that pops up without giving it much thought. In this environment, knowing how to ask the right questions - and dig through the answers - is a skill they sorely lack. And that's exactly what inquiry-based learning offers.

When we shift to the PEACE Framework approach, which blends AI with inquiry-based learning, we're giving students much needed critical thinking practice by challenging them to examine, evaluate, and even question AI-generated information. In other words, we build critical thinkers by asking them to become skeptical consumers of content. Instead of blindly accepting what they read or see, students learn to critique, to question, and to recognize the occasional nonsense that AI might spit out. And let's face it - in a world where misinformation is just a click away and conspiracy theories cower in every corner, that kind of critical thinking is priceless.

3. IBL Fosters Real-World Problem Solving

The world doesn't hand out clear-cut answers, and if we're honest, neither should we. Our students are stepping into a future where innovation and creativity aren't just bonuses - they're the core skills driving the workforce of tomorrow. The problem? In too many classrooms, "problem solving" is still about following a set of instructions to reach a single right answer that they can circle on the test. And sure, that might work for a textbook exercise, but it falls flat when it comes to solving real-world challenges.

Inquiry-based learning changes the game by embracing open-ended questions and real-world problems that require students to think beyond what's on the page. This model pushes students to get creative, think on their feet, and approach challenges without the safety net of a single "correct" answer. This ability to find innovative solutions to problems that don't come with instructions is a truly marketable skill for the workforce of tomorrow. A student who can ace a multiple-choice test on climate change is one thing, but a student who can think critically about renewable energy solutions and suggest ways their own community could benefit? That's something else entirely. The PEACE Framework uses AI-powered inquiry-based learning to focus on authentic problem solving because the future isn't about being ready for the next test - it's about being ready for the next challenge.

4. IBL Boosts Communication and Collaboration

Remember the days when students sat in rows, eyes front, pens poised to jot down endless notes? Those days are long gone. In today's world, communication and collaboration are king - and not just in the classroom. These are the top skills hiring managers are looking for right now, and they're only going to become more essential as AI reshapes the job market. But here's the catch: Thanks to the isolation of the pandemic, our students didn't just miss out on some academic content - they missed out on social skills, communication skills, and executive functioning skills that take years to develop. Now, we're not just teaching these skills; we're reteaching them and working to close a gap that widened into a chasm during remote learning.

Inquiry-based learning is uniquely suited to this task. By bringing students together to explore questions, discuss solutions, and tackle projects, it creates an environment where they're constantly practicing the kind of social interactions they missed out on. They're not just learning to take notes or complete assignments; they're learning to articulate ideas clearly, to listen actively, and to work through differences by integrating diverse perspectives into their work. Whether they're debating solutions, negotiating roles, or collaborating on a project, students are rebuilding those foundational skills they need for success in real-world settings.

And let's face it: Communication and collaboration are just as important as academic knowledge. Inquiry-based learning offers a safe space for students to practice these skills while relearning the art of collaboration, restoring the connection and shared purpose that was lost.

The world they're stepping into won't be one where they sit in rows, working alone. It's going to be one that requires empathy, flexibility, and teamwork - qualities that inquiry-based learning naturally cultivates, and the PEACE Framework supercharges with AI to boost academic skills, but also relearn what it means to be part of a team. Yes, in this new world AI will handle data and automate tasks, but humans still lead the way through connection and collaboration.

5. IBL Meets Diverse Learning Needs - Naturally

The fifth benefit is a big one, differentiation. That word can send even the most seasoned teacher's eyes rolling into the back of their heads. We loathe hearing this word because it's a monumental challenge to support students with a broad range of individual needs across multiple instructional grade levels with very diverse backgrounds and interests.

Meeting the needs of every student, all with their unique strengths, challenges, and experiences, is no easy feat. Our students are as diverse as they come. We're not just *one* type of learner; we're working with multi-language learners, students with learning disabilities, gifted and twice-exceptional learners, those with complex access needs, and even those dealing with challenges outside the classroom, like students who may be facing hunger, homelessness, difficult home environments, or the emotional impacts of blended and shifting families.

Inquiry-based learning rises to meet these challenges by creating a flexible, adaptable environment where differentiation isn't an afterthought, but built into the approach. By allowing students to ask questions and explore topics in ways that work for them, inquiry-based learning offers a personalized experience that goes beyond a one-size-fits-all curriculum. Students can work at their own pace and dive into material in ways that feel relevant and engaging.

For multi-language learners, this approach provides access to content through visuals, discussions, and hands-on activities that transcend language barriers. Students with learning disabilities or complex access needs can engage with content via videos, interactive media, and projects that allow for non-traditional demonstrations of learning. For gifted and twice-exceptional students, it provides the space to dig deeper into topics, ask challenging questions, and engage in higher-order thinking.

Inquiry-based learning goes even further by creating a classroom culture where students with non-academic issues, like unstable housing or food insecurity, can approach their work in ways that honor their circumstances. For some, that might mean working individually on a project they feel deeply about; for others, it might mean collaborating in small groups where they can build friendships and feel a sense of belonging. It's learning with flexibility, compassion, and respect.

And let's not forget the students who struggle with executive functioning skills, a challenge that has only been amplified for many since the days of remote learning. Inquiry-based learning is a powerful tool for rebuilding skills as students manage their own projects, set goals, and track progress. We teachers are there to offer guidance and scaffolding, but the autonomy and structure inherent in inquiry-based learning allow students to practice self-regulation, time management, and organization in a meaningful context, rather than as an abstract exercise.

Finally, the hands-on nature of inquiry-based learning also caters to various learning preferences. Whether students prefer text, video, audio, or physical experimentation, they can demonstrate their learning in a format that suits their strengths and interests. Some students may choose to present their findings through a multimedia project, while others might opt for a traditional report, a podcast, or even a creative performance. This flexibility allows every student to succeed in a way that feels personal and relevant to them.

To put a bow on it, inquiry-based learning doesn't just make room for diverse learners, it thrives on their diversity. By engaging students with content in ways that are meaningful to them, we're creating classrooms where every student has the chance to succeed, regardless of their starting point. And in a world that values creativity, problem solving, and adaptability, this approach isn't just ideal… It's essential.

Key Features of Traditional Inquiry-Based Learning

There are a few core features that make inquiry-based learning (IBL) shine. Let's connect the steps of traditional inquiry-based learning with the benefits:

- **Student-driven questions:** At the heart of IBL is the idea that students are asking the questions, not just answering them. Their questions propel the entire process, from orientation to discussion and reflection, which is why we have focused so heavily on questioning in prior chapters. This approach encourages students to take ownership of their learning, generate their own inquiries, and seek out answers through exploration and investigation. The benefit? Increased curiosity and a sense of empowerment, which are key ingredients for sustained engagement.
- **Active participation:** It's evident in all stages of IBL so students are not passive recipients of information. They aren't sitting back passively; they're in the thick of it by actively discussing, collaborating with peers, and diving into research.

Along the way, they're honing communication and teamwork skills. The benefit? Students are learning how to collaborate and share ideas, skills that will be crucial for their future careers and lives.

- **Process over product:** Unlike traditional learning models, where the end goal is the right answer, IBL focuses on the learning process itself. It's about reflecting on thinking, embracing trial and error, and developing the resilience that comes from learning through mistakes. It's about navigating the "messy middle". The benefit of this process-oriented approach? It fosters persistence and critical thinking, teaching students that the process of discovery is just as valuable as the final conclusions.

- **Real-world connections:** Finally, IBL doesn't happen in a vacuum. It connects academic concepts to the real world, making learning more relevant and meaningful. When students see how their knowledge applies outside of textbooks, they're more engaged and motivated to explore further. The benefit? Students can see how their learning fits within the world around them, and how the world around them can guide their learning, which feels purposeful and impactful.

Yet, as promising as traditional inquiry-based learning is, we've reached a critical moment for instructional pedagogy. The world is changing rapidly, and our educational approach has got to do some catching up to keep pace with technological advancements, particularly the rise of AI.

***

The Need for an Upgrade: Transition to PEACE

Think about toddlers for a second. You know the ones - they ask a million questions an hour, and half the time you're just trying to get through breakfast without losing your mind. "Why is the sky blue? Why do dogs bark? Why can't I have ice cream for breakfast?" It's endless. But here's the thing - we actually want that wonder in our world. Sure, it drives us a little crazy when they're small, but somewhere along the line, as kids get older, they start asking fewer and fewer questions. Part of that is because they're afraid of being wrong, or judged by their peers, or maybe school makes it feel like the right answer is more important than the right question. But squashing that curiosity? That's not the point of education.

Our goal shouldn't be to silence those questions just because they get inconvenient. We should be fanning the flames of that curiosity with inquiry-based learning. Now, if you're thinking that inquiry-based learning isn't exactly a shiny new concept, you're right. We've been talking about inquiry and exploration for decades.

91

Have you ever been teaching and started getting that feeling - like maybe you've been doing this the same way for too long? The world has changed, but our teaching methods seem to stay the same. And hey, I get it. We're all creatures of habit. Here's the thing, though - the classroom of the past, where knowledge was static and only available through textbooks and lectures, is no longer enough, and, honestly, traditional inquiry-based learning isn't enough either. But AI-powered inquiry-based learning? It's about to have its moment in the sun. Why? Because the way we approach instruction is fundamentally changing.

Accessing information is easy. What you do with information has become the real game. AI has essentially changed the rules of how we play, and if we keep teaching the way we did before, we're missing the point. Worst of all, we're doing a disservice to our students.

And that's where the PEACE Framework comes in. It builds on the solid foundation of traditional inquiry-based learning, but it also adapts to the possibilities and opportunities AI presents in the classroom. PEACE is the bridge between the classic methods that have worked well and the skills students need to thrive in an AI-driven world. We're not just tweaking what we already know; we're giving it a full upgrade, ensuring students can engage in inquiry in a way that's relevant for today's fast-paced, tech-driven reality. It's time to find peace with AI - so we can help our students go from knowing to *doing*, using what they know to innovate, create, and solve problems.

Here's a question for you: What happens when students don't just ask better questions, but use AI to answer questions that haven't even been asked yet? Stick around, because we're about to explore just that.

Chapter 4

PEACE: A 5-Step AI-Powered Inquiry-Based Learning Framework

Traditional inquiry-based learning brings a lot to the table, but as we step into the reality of AI-powered classrooms, it's clear we need to rethink the instructional pedagogy that guides our teaching. I get it; some of you might already be hitting the mental snooze button. I can almost see those glazed-over eyes. After all, it feels like every year there's another hurdle to jump, a new intervention to master, or another box to check. And if you're already in 'skim mode' because you've heard all this inquiry-based business before, let me stop you right there. *This time, it's different.*

Large language models and AI-generated content have changed education forever. Where we used to have students write extensively to gauge their
94

understanding and communication skills, a quick query with ChatGPT means we're no longer assessing what we think we're assessing. Simply put, a lot of our assessment go-to's no longer work when students can access AI in an instant.

But here's the thing - we don't need to throw out the baby with the bathwater. Instructional approaches like inquiry-based learning can still work, we just need to reimagine it for an AI-powered world. Think of it like upgrading from a flip phone to a smartphone. The basic function is the same (it's still a phone), but the capabilities are dramatically different. Remember flip phones? They made calls, sent texts, and maybe let you play Snake if you were lucky. Then, smartphones came along, and suddenly your "phone" became your camera, GPS, personal assistant, and entertainment center all rolled into one.

Did we stop making calls just because phones got smarter? Of course not. We just found better ways to communicate and use technology. The same goes for inquiry-based learning in the age of AI. We just need to find better ways to teach and use technology. Again, this is where the PEACE Framework comes in.

Instead of fighting against the AI tide or pretending it doesn't exist, PEACE leans into AI as a tool for deeper learning. It's not just about enhancing inquiry-based models... It's about rethinking our entire instructional approach. This five-step system - **Provoke, Enquire, Analyze, Create, and Engage** - transforms how we approach student questioning in an AI-powered classroom.

The beauty of **PEACE** is that it doesn't just work around AI - it works with it. Each phase intentionally incorporates AI as a tool while maintaining the critical thinking and creativity that make inquiry-based learning so valuable. It's like giving students a power-up in a video game: The fundamental skills are still essential, but now they have enhanced capabilities to explore, create, and learn.

Let's take a quick tour of **PEACE Framework** before we dive into the nitty-gritty, like your GPS overview before we zoom in street by street.

PEACE starts with **Provoke**, where we spark curiosity in our students. You know that moment when a kid's eyes light up because something finally clicks? That's what we're after. But instead of just hoping it happens, we're going to engineer it.

Then, we move to **Enquire**, where we teach students to ask better questions. Not just surface-level Google-able stuff, but the kind of questions that even AI might struggle with. (And yes, I did spell enquire with an 'E' on purpose - we'll get to why later.)

In the **Analyze** phase, we're teaching students to be AI-savvy investigators. We're upgrading their research skills from "copy and paste" to "evaluate and iterate." This is where prompt engineering becomes their power up, not their crutch.

Create is where students take everything they've learned and actually make something with it. And no, I don't mean having AI generate a presentation while they play Minecraft. We're talking about genuine creation that showcases their understanding.

Finally, **Engage** is all about sharing and teaching others. Because you never really know something until you have to explain it to someone else. (Every teacher reading this just nodded in agreement.)

You might be thinking, "Great. PEACE is just another edu-buzzword." But stick with me here, because the beauty of this framework is that these phases flow together naturally, just like a good lesson should. Each phase builds on the last, but they're also flexible enough to revisit and adjust as needed. Because, if there's one thing we know about teaching, it's that no plan survives first contact with actual students.

Let's unpack each phase starting with *Provoke*, where we'll explore how to ignite wonder in a world where answers are just a chat prompt away...

***

Provoke

We've already established that questioning is essential. Chapter 1 unpacked the shift from memorization to mastery, spotlighting the role of thoughtful questioning in modern education. Chapter 3 dove into inquiry-based learning, highlighting how curiosity can drive critical thinking. Now, it's time to zoom in on the first phase of the PEACE Framework: *Provoke*.

This is where everything starts... By igniting curiosity and setting the stage for deeper exploration. But *Provoke* isn't just about getting students to ask questions; it's about cultivating the kind of questions that spark excitement, challenge assumptions, and dig into complexities.

Think of it as striking the match that lights the fire. This phase is all about creating moments where students don't just *engage*, they lean in with wide-eyed wonder. It's less about teaching the value of questioning (we've covered that) and more about building the kind of environment where curiosity becomes contagious.

There's something to be said about sparking curiosity within others.

My son, Luke, came to me one day with a piece of paper, a ruler, and his wallet. "Wanna see this?" To set up his 'magic trick', he placed the ruler on the edge of the table, about half on, half off. Then, he held the wallet high over the edge of the ruler and asked me what was going to happen. I told him the ruler was about to pop him in the face. Luke dropped it and the ruler went flying. The commotion drew his little brother, Leo, into the room.

Then, Luke set the ruler up again. This time he wanted to bet me. He told me I could keep all the money in his wallet if his ruler hit the ground, BUT he got extra tech time if it didn't. How could I resist? He put the ruler on the table again, half on and half off, but this time he placed a single piece of paper over the ruler, just on the half that was on the table. Leo leaned in and yelled that the paper would never be able to keep the ruler from flying after the wallet hit it. He was primed for an epic fail. A Cheshire smile appeared on Luke's face, and he dropped the wallet a second time.

The ruler never fell. Luke got his extra tech time. Mission accomplished. Meanwhile, Leo was beyond amazed. He couldn't believe what had happened. He wanted to know why it worked, if a heavier thing would change the odds, if there was a tipping point for the ruler to balance on the table. Literally, a hundred questions and hypotheses barreled out.

That is the power of provoking thought; curiosity unleashed.

See what happened there? A simple ruler trick didn't just catch Leo's attention - it launched him into full inquiry mode. He wasn't asking these questions because a worksheet told him to. He wasn't generating hypotheses because it was assignment number three on the rubric. He was genuinely curious. That's what we're after in the Provoke phase.

But here's where it gets interesting in our AI-enhanced classroom. Where we used to worry about students finding answers too quickly, we can now use AI to help us create even more intriguing provocations. Think about it - AI can help us generate hundreds of "what if" scenarios, create compelling discussion starters, and even help us identify the perfect entry point for different types of learners that sparks curiosity.

The trick (pun intended) is to create 'ruler moments' that make students forget about getting the "right" answer and focus instead on their own curiosity. This pulls forward all the things from Chapter 1 and the power of asking better questions (which we'll discuss more at the *Enquire* phase). Leo wasn't satisfied with knowing the paper trick worked - he wondered about the physics behind it, how he could test its limits, and ways to explore variations. *Provoke* is the cornerstone of PEACE because it ignites curiosity as a pivotal approach to learning. This kind of curiosity doesn't happen by accident; it's sparked through intentional strategies that invite students to wonder, experiment, and innovate.

This phase focuses on generating wonder and giving students a clear pathway towards innovative thinking, such as:

- **Thought-Provoking Scenarios**: Present real-world dilemmas or futuristic what-ifs that demand curiosity.
- **AI-Powered Brainstorming**: Use AI to provide surprising facts, alternative perspectives, or prompts that challenge students' assumptions.

- **Collaborative Curiosity Builders**: Pair students to generate questions in response to open-ended AI outputs or visual prompts.

Let's break down how to create these 'ruler moments' in your classroom...

Connect to Prior Knowledge

Have you ever spent an entire weekend crafting the *perfect* lesson plan, to realize Monday morning that half the class was lost before you even got past the intro? We've all been there. It's the classic teacher's dilemma: How do we bridge the gap between our carefully planned lessons and our students' diverse backgrounds?

Each student comes to us with a unique blend of experiences, knowledge, and abilities, so we have to start by tapping into what they already know. Think of it like building a house: You wouldn't set a skyscraper on shaky ground. Likewise, we can't expect students to grasp complex concepts without giving them a solid foundation. But let's be real… laying that groundwork isn't always easy. We're all racing against the clock to cover our standards, and sometimes it feels like we don't have a moment to spare. But skipping this step?

103

That's like cutting corners on the foundation, and it'll lead to some serious collapses later on.

So, let's look at a few *quick wins* to activate prior knowledge effectively and get everyone on the same page:

- **Quick Write and Share**: Start by asking students to jot down everything they know about the topic in a few minutes. Then, have them share their thoughts with a partner or group. If tech is available, use a shared document to make it a live brainstorm, letting students see each other's ideas unfold. This helps them organize their thoughts and spot any knowledge gaps.
- **Verbal Brainstorm**: Skip the paper and go straight to sharing out loud. Invite students to call out what they know in a rapid-fire, informal way within a whole or small groups. This keeps things quick and lively, getting everyone involved without needing to write.
- **Share Media**: Show a picture, video, or graphic that relates to the topic. Then, have students break into small groups to discuss what stood out. Visuals are a great way to stimulate thinking and

curiosity, giving students something concrete to connect with before diving deeper.

- **Anticipation Guide**: Provide a list of statements (true, false, or opinion) related to the topic and ask students to agree or disagree before diving into the material. After the lesson, revisit their answers to reflect on any shifts in understanding. This list is easy to create with your favorite AI assistant.

These strategies have big payoffs. First, they activate communication skills through peer-to-peer interactions, making for an engaging start. And even better, students learn from each other's experiences, often filling in gaps they didn't know they had. It's a powerful way to bridge that tricky prior knowledge gap.

Here's a quick example: Let's say you're starting a unit on the water cycle. Begin by asking students to share their experiences with water - whether it's drinking water, swimming, or even the time they got caught in the rain without an umbrella (we've all been there). This kind of conversation gets the gears turning and primes them for what's next. Plus, it's far more effective than kicking things off with a lecture or, heaven forbid, a reading assignment. Don't worry, we'll get to the reading and comprehension skills later in the process; for now, we're looking for a quick win and a fast way to establish a shared foundation.

By taking the time to connect with prior knowledge, we're not just setting up the lesson; we're creating a launchpad for inquiry. This foundation will support students as they dive into more complex questions and tasks in the PEACE Framework - ultimately helping them build confidence and engagement right from the start.

Spark Curiosity

You know what's funny? We do this all the time in our own lives. Think about your doom-scrolling habits. (Come on, we all do it.) When do you actually stop mid-scroll? When something seems improbable, unlikely, or just plain weird. That's what makes you pause and think:

Is it real? How does that work? Could I do that?

Just like Leo with the ruler trick, our human instinct is to be curious when things seem unusual. The good news? We can harness this same instinct in our classrooms. Even better news? AI can help us do it without spending hours hunting for the perfect 'ruler moment'.

Tap into things that are surprising and weird or that challenge students' assumptions and encourage them to think critically. Present unexpected results, interesting data sets, sensational news headlines, or even social media reels to prompt question-asking. And use AI to cut your workload and also spark this curiosity in a way that aligns with grade-level standards and skills.

Here's a practical example: Instead of just telling students about animal adaptations, I asked an AI to generate some curiosity-provoking statements. One of the first responses was "Some pigs can swim better than humans." I wish you could have seen my students' faces. Just like Leo with the ruler, they had a hundred questions. Why? How? Which breeds? Won't they just sink? How fast can they swim?

But here's where we need to be strategic. It's not just about finding cool facts - it's about creating moments that connect to our learning objectives. When I'm planning these provocations, I use this simple prompt with my LLM:

107

> "Generate a list of 10 curiosity-provoking statements with unexpected results or interesting information. Align them with my [grade level] lesson about [topic or objective]. Pair each statement with a curiosity/thought-provoking question."

Now, we've gone beyond just using AI to make some cutesy animal-themed worksheets. We've used its limitless knowledge to provoke our students into considering possibilities outside their frame of reference. In doing so, we can move them towards the grade-level standards in a way that promotes all the benefits of inquiry-based learning we discussed earlier.

Because I can't help myself, I have to offer out the other fascinating result of my curiosity driven query: Giraffes have the longest necks of any land mammal, but they only have seven vertebrae in their necks- just like humans. Ruler moment?

See, I've provoked you too, haven't I?

108

Classroom Vignette: Generating Wonder with AI

Ms. Bechi's 4th-grade classroom falls silent as an intricate 3D model of a honeycomb appears on the screen. What began as a standard lesson on geometry has transformed into an exploration of nature's mathematical precision.

The Setup

Ms. Bechi had asked Claude to generate increasingly complex examples of hexagons in nature, each with an unexpected mathematical connection. The honeycomb visualization included measurements showing how bees maximize storage space while minimizing building materials.

Wonder in Action

"Who can tell me why bees might choose this shape?" Ms. Bechi asks, watching her students' eyes widen.

Ara raises her hand hesitantly. "Is it... because hexagons fit together without gaps?"

"Excellent observation! What else do you wonder? Get with your elbow partner and Question Storm."

The questions begin flowing:

- "Do other insects build hexagons too?"
- "How do bees measure the angles so perfectly?"
- "Could we build houses like this to save space?"

AI Integration Points

Morning Preparation: Ms. Bechi used AI to generate grade-appropriate fun facts connecting geometry to nature.

During Class: Students suggest objects to analyze for geometric patterns.

Extension: AI helps generate differentiated investigation paths based on student interests.

Sample Learning Journey: Marcus's Discovery Path

1. **Initial Question:** Bee hives are not the only place to find shapes in the world around us. What are examples of geometry in nature?

2. **Marcus's Wonder:** "Why do snowflakes have six sides?"

110

3. **Marcus's AI -based Findings:** Snowflakes are also hexagons!

4. **Project:** Creating his own symmetrical designs inspired by snowflakes and hexagons.

Marcus's Journal Entry:

"I never knew math could be so cool! When Ms. Bechi showed us the honeycomb thing, I was like 'whoa!' I thought hexagons were just shapes in our math book, but they're everywhere! My favorite part was when I got to design my own snowflake pattern. The AI helper gave me ideas about symmetry, but I made the final design all by myself. Now I keep seeing shapes everywhere I look, even in the cereal I eat for breakfast! My little sister got annoyed because I wouldn't stop talking about hexagons at breakfast. ☺"

Teacher Reflection

"The key was using AI to create that initial 'ruler moment'. Before, I used to use a lesson on stop signs. Now, that seems like a total bore! This question-driven way really challenged assumptions, made math interesting, and allowed curiosity to take over. Once they saw the honeycomb visualization paired with the curiosity question, their questions became more sophisticated... and there were so many! AI helped me transform a standard geometry lesson into an investigation of patterns in nature with a touch of engineering and even cultural design. The technology helped me tap into the wonder and get students engaged, but curiosity drove the learning."

Ask Open-Ended Questions

So, we've activated their prior knowledge and sparked their curiosity. They're buzzing with ideas, questions, and maybe even a few wild tangents. The challenge? All that energy can be hard to harness. Left unchecked, curiosity can send students in a thousand different directions - which is great for sparking interest, but not so ideal when you're trying to reach a specific learning objective.

That's where open-ended questions come in, and here's the key: They don't just encourage exploration - They channel it. Imagine a classroom where students are asking, "Why do plants need sunlight?" "Can plants grow on Mars?" "How does photosynthesis

112

work?" and "Can a plant grow without water?" Instead of a jumble of disconnected questions, our goal is to focus their curiosity toward the lesson's objectives. When we ask open-ended questions that guide students toward specific grade-level standards, we're not only teaching them to inquire but also building a bridge between their natural curiosity and our instructional targets.

Here's an example: Rather than starting with, "What is photosynthesis?" (a question that leads straight to a definition) we might ask, "How do plants adapt to different environments to survive?" This question is open enough to let students think broadly, but it also points them toward specific concepts - adaptation, survival, and the relationship between organisms and their environment. Now, instead of running off in every direction (or to Mars), their curiosity has a directional spark, aligning naturally with the day's learning goals.

And here's where AI shows up to be our helpful teaching assistant. Ask your favorite LLM to generate a set of open-ended questions that relate specifically to your standards. Let's say your goal is to explore ecosystems. Ask, "Generate a list of open-ended questions about how animals interact within ecosystems." Now you have questions that fuel curiosity but don't stray too far off the intended path. By using AI strategically, we guide students toward a productive inquiry that builds toward core knowledge rather than veering into the weeds.

To get those gears turning and zero in on learning targets, here are some ways to approach open-ended questioning:

- **Frame the Big Picture**: If your lesson is about adaptations, instead of defining terms, start with, "Why do some animals survive better in deserts than others?" This keeps curiosity focused on adaptations while leading them naturally to the concept.

- **Guide Curiosity with Constraints**: Instead of asking students what they know about weather, try "What might happen to our environment if rainfall patterns changed?" This still engages them but within the context of your instructional goal.

- **Encourage Process-Oriented Thinking**: When exploring food chains, ask, "What would happen if one species in the food chain disappeared?"

This invites them to apply critical thinking specifically within the context of ecosystems.

If you want to try this with your grade level, content focus, or topic, here are two prompts to feed to your LLM teaching assistant:

"Generate a list of 10 open-ended questions about (lesson objective or content focus) to use for inquiry-based learning with (grade level) students."

"Generate essential questions to guide (grade level) grade student inquiry throughout a lesson on (topic/content)."

By using open-ended questions to channel curiosity, we're setting the stage for students to engage with intention. They're learning not just to ask questions but to dig into the kinds of questions that bring them closer to understanding the day's learning objectives. With this approach, curiosity becomes a targeted tool that helps them connect with complex concepts on a deeper level.

Examples of Open-Ended Questions

Before we move on, I want to give you the opportunity to see more open-ended questions in action as well as how they can relate to grade level standards and skills and multiple instructional levels.

Here are a few examples to help you see the power of open-ended questions.

Elementary School Examples

- Instead of asking, "What color is the sky?", try "What are some different colors you've seen in the sky?" *(Writing Standard: Developing narrative and informational writing, using descriptive language, and organizing ideas.)*

- Rather than "Who is the main character in the story?", ask "How would you describe the main character?" *(Reading Standard: Analyzing text, making inferences, and supporting claims with evidence.)*

- Ditch "What is 2+2?", and go with "Can you create a story problem where the answer is 4?" *(Number and Operations Standard: Applying mathematical concepts to real-world problems, developing number sense, and using mathematical reasoning.)*

Secondary School Examples

116

- Instead of "What caused the American Civil War?", try "What were the different perspectives on the causes of the American Civil War?" *(History Standard: Analyzing historical sources, interpreting historical events, and evaluating historical arguments.)*

- Rather than "What is the Pythagorean theorem?", ask "How can the Pythagorean theorem be applied in real-world situations?" *(Geometry Standard: Proving theorems, using geometric properties, and solving problems involving measurement.)*

- Ditch the "What are the different types of renewable energy?", and go with "How can we create a more sustainable energy future for our community?" *(Science Standard: Understanding biological processes, ecosystems, and the diversity of life.)*

Clearly, open-ended questions are a key ingredient to success at the Provoke phase. They lay the groundwork for inquiry that's both expansive and purposeful. However, that's not enough to get us where we need to be by the end of this phase. What if you could amplify the power of your open-ended questions and further encourage students to stretch their thinking beyond the basics? That's where thought-provoking questions come into play!

Add Thought-Provoking Questions

Open-ended questions are essential to the Provoke phase... they open the door to inquiry. But what if we could take it a step further and add fuel to the fire? Thought-provoking questions do just that, pushing students beyond basic inquiry to deeper levels of reflection, imagination, and critical thinking. These questions are surprising, unusual, even a little controversial, and they challenge students' assumptions, encouraging them to see the topic from different angles.

Now, here's the best part: With AI as a thought partner, creating these questions is easier than ever. While the teacher down the hall is using AI for generating worksheets, you're taking it to the next level with AI as a collaborator in designing questions that get students genuinely excited. You can even task AI tools to gather multimedia, recent events, or real-world scenarios that align directly with your standards-based objectives. The magic of AI here is that it doesn't just help you create questions - it helps you craft questions that are both curiosity-driven and standards-aligned.

Here's a sample prompt for generating these questions with AI:

"Generate a list of 5 curiosity-based questions for inquiry-based learning with (grade level) students based on (topic) who are learning about (curriculum outcomes or lesson objectives)."

I tried this with an LLM for a 3rd-grade unit on the water cycle, using the prompt:

"Generate a list of 5 curiosity-based questions for inquiry-based learning with 3rd-grade students based on the water cycle who are learning about condensation, evaporation, etc."

Here were the results:

- What do you think happens to a puddle of water on a sunny day? Where does the water go?

    - *This question introduces the concept of evaporation in a relatable way.*

119

- Why do clouds look fluffy and white sometimes but gray and heavy at other times? What do you think is happening inside the clouds?

  - *This question leads to discussions about cloud formation and the role of condensation.*

- When it rains, where do you think the water goes after it falls to the ground?

  - *This prompts students to consider the various paths water takes after precipitation.*

- How do plants use water from the soil, and what happens to that water when it's inside the plant?

  - *This question connects the water cycle to the process of transpiration in plants.*

- If you could create a mini water cycle in a bottle, what steps would you include to make it work?

  - *This encourages creativity and critical thinking about the components of the water cycle.*

I love these questions! I could easily turn this list into a whole unit (or even a summer camp)! And here's why: Even as someone who knows the science behind the water cycle, these questions make me pause, wonder, and see the topic with fresh curiosity. That's the power of thought-provoking questions - they aren't just about reviewing facts; they invite students to think critically, dream a little, and connect with concepts on a deeper level.

By leveling up our open-ended inquiries with thought-provoking questions, we set up a highly engaging learning environment that encourages curiosity to thrive within clear, purposeful boundaries. And, as we'll see throughout the PEACE Framework, these questions don't just kick-start inquiry; they keep it going, acting as a guide to help students dig deeper, make connections, and move closer to a nuanced understanding of complex topics.

Brainstorm Questions

Remember that scene in "Inside Out" where memories start popping up like popcorn? That's exactly what we want happening with questions in the Provoke phase. But how do we get there?

Let me share what's worked best in my classroom. We start with what I call a "Question Storm". (I learned the hard way not to call it a 'question dump' after my middle schoolers had a field day with it - you're welcome for that pro tip.) The goal is simple: Get every single question out there, no matter how basic or brilliant. Think of it as creating a question buffet - we'll worry about nutrition later.

P.S. I'm going to continue to use this phrase 'question storm' to refer to brainstorming questions. Just an FYI.

Question Stems: Your Provocation Power Tools

Sometimes students stare at you like deer in headlights when asked to question storm. That's where question stems come in. They're like training wheels for curious minds. Here's your starter pack:

- "How might _____ change if _____?"
- "What would happen to _____ if _____?"
- "Why does _____ matter to _____?"
- "How does _____ connect to _____?"

Think of these stems as kindling for your curiosity fire. They give students just enough structure to spark their own questions without boxing them in. And here's the beauty part… you can adjust them based on your students' needs and your lesson objectives.

Interested in getting more question stems to use in your classroom? You can grab 20 printable stems at www.peaceframework.com/provokestems

Special Education Spotlight: Visual question builders work wonders here. Create a set of laminated cards with different question stems. Write some of the key vocabulary from your unit on sticky notes. Now, students can physically mix and match to build their questions. It's like magnetic poetry, but for inquiry!

The goal of this brainstorming isn't just to generate a pile of questions… it's to prime the pump for deeper inquiry. Think of it as laying the groundwork for the more focused questioning we'll do in the Enquire phase. We're not just teaching students to answer questions; we're teaching them to become question connoisseurs.

Remember: In the Provoke phase, there's no such thing as a bad question. We're creating an environment where curiosity runs wild and every "I wonder..." gets its moment to shine. The filtering and refining? That comes next. For now, let's just get those questions flowing!

In a Nutshell: The Provoke Phase

Here's how the Provoke phase flows in practice: We start by building on prior knowledge, move on to sparking curiosity, level up by layering in a thought-provoking, open-ended question, and then finish by letting our students loose to brainstorm what they wonder about. Think of it as a progression that guides students from what they know to what they wonder and, ultimately, to what they're excited to learn.

Here's an example to bring this to life:

"Adaptations are the changes animals make to survive - whether that's becoming a better predator or learning to avoid becoming prey."

"What do you already know about animal adaptations? Turn to your elbow partner*, and each of you take 30 seconds to share everything you know."

"Now, what if I told you that some pigs can swim better than humans? Why do you think that is? Take a minute with your face partner* and see what ideas you can come up with."

"Alright, as we dig into animal adaptations, here's my question for you: How do animals that live in the desert find water to drink?"

"Take a few minutes to wonder about this and then question storm."

*Your elbow partner is sitting next to you. Your face partner is sitting across from you.

In this sequence, the Provoke phase builds excitement and sets students on a path of meaningful inquiry. Each step builds on the last, moving students from their initial knowledge to a more complex understanding.

Remember when we talked about engineering those 'ruler moments' where students are brimming with curiosity and wonder? That's exactly what the Provoke phase is all about. It's where we set the stage for everything that follows in our PEACE Framework. Here's how it all comes together:

- **Activate prior knowledge**: Think of this as warming up your students' mental engines. We're connecting new ideas to their existing mental map, making the unfamiliar feel a bit more familiar.
- **Spark curiosity**: This is where we throw in our unexpected twists, our "wait, what?" moments that make students sit up and take notice. Whether it's a paradox, a provocative statement, or an unusual demonstration, we're aiming to crack open their curiosity.
- **Plant purposeful provocations**: These strategic prompts align with our learning objectives but leave room for wonder. These open-ended thought-provoking questions are like breadcrumbs leading students toward deeper inquiry.
- **Launch the question storm**: Now that we've got their mental gears turning, we let their curiosity run wild. Every question is welcome - from the seemingly simple to the surprisingly

sophisticated. We're building a question bank that we'll mine for gold in the Enquire phase.

Remember, a good Provoke phase doesn't just capture attention; it creates an itch that students can't help but scratch. When you see them leaving class still asking questions, that's how you know you've nailed it.

When we're intentional about provoking curiosity, we're not just teaching facts. We're creating a learning environment where students are eager to dig deeper, ask questions, and follow the trail to their own discoveries - setting the stage for the next steps in the PEACE Framework.

Next up: The Enquire phase, where we turn curiosity into thoughtfully crafted questions.

***

Enquire

Remember that time your English teacher marked your paper and corrected "enquire" to "inquire"? Or maybe it was the other way around? Well, here's a fun fact that might make you feel better: Both spellings are correct, but they're not quite the same thing. "Inquire" is about conducting a formal investigation - think detective work or scientific research. But "enquire"? That's more about asking questions out of genuine curiosity, like when a student's eyes light up and they just *have* to know more.

That's exactly why I chose "Enquire" for this phase of the PEACE Framework. We're not turning our students into private investigators (though that might make for an interesting lesson plan). We're nurturing their natural curiosity, teaching them to ask questions they are curious about. In today's AI-powered classroom, that matters more than ever. So, from here on out, it's enquiry rather than inquiry!

Have you ever wondered why questions are so powerful? They're not just for gathering facts - they're what drives learning and sparks critical thinking. Let's break down what a solid, well-crafted question can do.

First up, questions light a fire under curiosity. We've talked about this before, but it's worth repeating. A well-crafted question can spark interest, encouraging students to dive deeper, even into topics they'd normally gloss over. With the right open-ended, curiosity-driven question, we get students to think critically and dig deeper into the topic at hand.

Second, quality questions push for deeper understanding. They help students uncover the nuances and complexities in a subject, moving beyond just surface-level facts. Think of a student studying the American Civil War; they might ask, "What were the economic differences between the North and South?" or "How did the abolitionist movement impact the war?" These kinds of questions push students to dig into the historical context and develop a richer, more nuanced understanding of the events.

Third, questions trigger critical thinking, something facts alone struggle to do. Asking quality questions helps students analyze information, assess evidence, and form their own opinions. Just like after reading a persuasive essay, students might be asked to identify the author's main argument, evaluate the supporting evidence, and consider counterarguments. This process builds those critical thinking muscles they'll need both in school and in life.

Finally, questions drive problem solving. When faced with a challenge, asking the right questions helps students break down the problem, identify key factors, and come up with solutions. For example, if someone was tasked with designing a sustainable energy solution for their community, they might ask questions like "What are the most common sources of energy in our area?" or "What are the environmental and economic benefits of different renewable energy options?" These real-world problem-solving skills are exactly what students need for the AI-powered workplace. So, by asking quality questions, students can learn to identify the key factors to consider and develop a well-thought-out solution.

As part of this Enquire phase, we want to get our students to question, but how do we do that?

Guiding Students to Ask Quality Questions

Last week, I watched my little sister tackle her history homework. Like any tech-savvy 16-year-old, her first instinct was to whip out her phone and ask ChatGPT about the causes of World War I.

"Why would I read it when AI can answer for you?" she grinned.

I couldn't help but smile, remembering my own school days when the cutting-edge technology was asking Jeeves (who wasn't nearly as helpful as he claimed to be). But what happened next was fascinating.

ChatGPT gave her a perfectly reasonable answer with a neat, bulleted list of causes. My sister started to copy it down, but then paused. "Wait," she said, frowning at her phone, "what does 'militarism' actually mean? And how did all these things work together?"

That's when it clicked. Without realizing it, she'd stumbled into real learning. Her initial question led to more questions, each one deeper than the last. By the end of the conversation, she wasn't just copying facts about WWI; she was genuinely exploring how complex international relationships and human nature itself can spiral into conflict.

Here's the thing: Left to their own devices, most students will do exactly what my sister first tried to do

- ask AI for an answer and stop there. They'll miss the real power of these tools, which isn't in getting answers, but in sparking better questions.

That's why we can't just throw our students into the AI deep end and hope they'll figure out how to swim. The harsh reality? They won't. We need to guide them through the art of questioning, show them how to query effectively, and teach them to iterate their conversations with AI. This is much different than a quick splash of a Google search in the shallow end of fast answers. It's about learning with AI through deep dives, exploring our understanding from multiple angles, and becoming confident swimmers in these deeper intellectual waters.

This is where the real work of the Enquire phase begins. It gives us a structured approach that helps students develop these critical questioning skills. By sharpening their questioning protocols, modeling effective queries, using think-alouds, providing question stems, and promoting peer collaboration, we can transform our students from passive AI consumers into active enquiry leaders.

BID Question Protocol

Let's be real - students come up with all sorts of questions, especially in a free-thinking phase. Just last year, I had a student who asked if dolphins could survive in a forest (spoiler: they can't). But that question sparked something real inside me about why questions are powerful, especially when they lead to deeper enquiry. Imagine if we could channel that natural curiosity into something more structured, something that really digs beneath the surface.

The BID Routine

That's where question protocols like the BID routine come in. Think of them as structured routines that help students frame their curiosity, moving from rough ideas to thoughtful questions that spark deeper thinking. The BID routine (**B**rainstorm, **I**dentify, **D**raft) is a flexible approach that can help bring order to the early stages of questioning. It works seamlessly as a bridge from the Provoke phase into the Enquire phase, helping students to refine, reflect, and reiterate.

Here's how the BID routine works in practice:

- **B - Brainstorm:** Students start their BID where they left off in the Provoke phase - with a question

133

storm. It's our bridge from Provoke to Enquire. By brainstorming as many questions as they can, they have a chance to explore all their curiosities and ideas, no matter how big, small, or even unrelated. No filters, no judgments; just ask. This free-flow approach lets students open up without worrying about whether their question is "good" yet.

- **I - Identify:** After generating a wide pool of questions, students move on to identifying the ones with the most potential. Here, they sift through their brainstorm and select a few key questions that seem promising, are open-ended, or spark interest. They also bounce their ideas off each other through collaboration. This step is about finding those "strong bids" for enquiry; the questions most likely to uncover valuable insights or be most interesting.

- **D - Draft:** In this final phase, students draft (or re-draft) their top questions. They work on refining them to be as clear, purposeful, and as specific as possible by adding the elements that will drive deeper enquiry. This is where they prepare their "bid" by honing their questions to be insightful, meaningful, and ready for focused exploration.

And yes, as you may have guessed, the BID routine is a hidden collaboration exercise. Students discuss their selections, debate which questions are most valuable, and share their reasoning. It's an excellent way for them to see how different perspectives shape enquiry. Plus, it builds essential skills like critical thinking and teamwork along the way.

With **BID**, students aren't just asking questions; they're learning that each question is an "opening bid" that can be improved with every round. Through second and third bids, students discover the power of iteration, seeing how each refinement strengthens their enquiry and sharpens their focus. It's a dynamic routine that builds curiosity, supports critical thinking, and encourages students to view questioning as an *evolving process.* The key is helping students understand that their first bid is just that - a first attempt.

The ReBID Process

Now that your students have explored their initial BID questions, they may have discovered it was not exactly the question they wanted to ask or maybe it wasn't as focused as they had hoped. It's time to ReBID! Just like in a real auction, students take their initial "bid" and make it even stronger.

- **RE - Review & Evaluate:** Review the final draft from the last BID cycle. What were its strengths and weaknesses?
- **B - Brainstorm:** Using your last question as a starting point, question storm by considering new ideas or perspectives.
- **I - Identify:** Select the most promising one and focus on ways to refine and reshape the question.
- **D - Draft:** Create a new, improved question that incorporates your refinements.

Each pass through the BID routine is an opportunity to refine, reshape, and ultimately craft questions that drive deeper learning. So, encourage them to keep bidding until they hit that sweet spot where their questions unlock the understanding they're looking for.

You may be asking yourself why we spend so much time on questioning. It's because answers are easy to find in our AI-powered world. In order to adapt instruction, we need to shift our focus from finding answers to the art of asking. By going through a BID and ReBID routine, students think. In that thinking, they are learning how to be problem solvers, critical thinkers, and wordsmiths. *The learning is in the questioning,* which is why this routine is such a powerful part of the PEACE Framework.

BID Routine Alternatives

Now, for days when you want a different approach than the BID routine, there's also the **5 Whys** method, which, admittedly, is a bit like channeling your inner toddler. (If you've ever been asked, "But why?" for the fifth time in a row, you know the drill.) By asking "why" five times, students get to the root cause or primary reason behind something. Imagine they start with, "Why does the water cycle matter?" By the fifth why, they're typically in much deeper territory, like discussing ecosystems and climate stability - questions that connect science to the broader world they live in.

Here is an example of the 5 Whys in action:

1st Why: Why does the water cycle matter? **Answer**: Because it helps to circulate water around the Earth.

2nd Why: Why do we need to circulate water around the   Earth?

Answer: Because water is a finite resource on the planet.

3rd Why: Why do people waste water if it's a limited resource?
Answer: Because they don't realize how hard it is to transform water into usable forms.

4th Why: Why don't people know how water moves on the planet?
Answer: Because they don't think the water cycle applies to them.

5th Why: Why don't we teach people how the water cycle impacts the drinking water they have in their house?

Now we have a great question to work from! It has evolved into something much deeper and more meaningful - one that prompts both action and reflection. This question highlights the importance of water conservation and education in a way that resonates with students, connecting directly to their inner thought process and real-world concerns.

Or maybe you prefer a visual approach, in which case the **Fishbone Diagram** might be helpful. It's perfect for students who think best by seeing connections mapped out. Students start with a central question, then "branch out" into different causes and effects related to it. It looks a bit like a fish skeleton (hence the name), and while students find that a little amusing, it's highly effective for identifying variables and potential impacts.

Using thinking routines can also be a helpful way to visualize thinking and cultivate questioning. If you'd like to learn how to incorporate thinking routines in your classroom and boost enquiry-based learning, join me for a quick session here at www.peaceframework.com/thinkingroutines

---

And here's one more way to dig deeper - the **5 W's and H** method. This is all about covering the bases - **Who, What, Where, When, Why, and How**. After students generate their initial questions, this protocol helps them flesh out each one by considering different perspectives and contexts. Say a student asks, "How do animals adapt to their environments?" They can then break that question down into the W's and H:

- **Who** does this affect (which animals, or which specific populations)?
- **What** types of adaptations are we looking at - physical, behavioral, or both?
- **Where** does this adaptation occur, and is location a factor?
- **When** did this adaptation likely start, or how quickly does it occur?
- **Why** does this adaptation matter to the species or ecosystem?
- **How** does it actually happen - what mechanisms drive it?

This process of implementing a question protocol is about teaching students to examine each question they've generated to see if it's worth investigating or simply interesting to know. Suddenly, they're turning curiosity into focused and actionable enquiries.

Using structured questioning protocols is like giving students a roadmap... they can take scenic routes and explore side streets, but they won't end up completely lost in the wilderness of their wonderings. And best of all, these protocols teach students that questions aren't just about getting to a destination or about finding different paths to a singular right answer. Instead, these protocols become the rocket fuel for launching into deeper thinking and better prompt engineering.

When all else fails, let AI help refine your questioning! Start by inputting a general question, but before hitting enter, add this phrase: *"Before returning your results, generate three clarifying questions to help me focus my enquiry."* This simple tweak encourages AI to support your critical thinking process by narrowing broad topics into more actionable and precise directions. It's a practical way to use AI as a partner in crafting better questions and driving deeper enquiry.

Drafting Better Questions with Elements

Let's pause for a second and remember how we got here. First, during the Provoke stage, we activated prior knowledge with some collaborative pair-shares. Then, we sparked some curiosity and pivoted to a thought-provoking question. Our students were jazzed up, and we had them wonder before collaborating again to share their ideas and maybe spark a few more insights. Next, we guided students in shifting from wonder questions to enquiry questions by implementing structured protocol, like the BID and

ReBID routines, helping them dig deeper into their initial wonderings.

But what if their questions are not that great? What if they BID and ReBID, but still miss the mark? We have to teach our students how to draft better questions by adding focus in a way that propels them toward deeper enquiry.

Start with Simple Questions and Gradually Increase Complexity

Drafting great questions doesn't mean dismissing the simple ones. Those are the foundation. Think of it like cooking a meal: You start with basic ingredients and gradually layer in flavors to create a rich, complex dish. Each ingredient matters. For our "question recipe", the ingredients are the elements we add to questions to make them more robust.

Here's an example to show how a question can evolve using elements:

- **Start**: "What are the causes of the American Civil War?"
- **Add a Temporal Element:** "What were the major events and developments in the decades leading up to the American Civil War?"

142

- **Add a Perspective Element:** "What were the key economic and social factors in the decades leading up to the American Civil War?
- **Add a Comparative Element:** "Compare and contrast the economic and social conditions in the Northern and Southern states in the decades leading up to the American Civil War."
- **Finished Prompt:** "Explain the economic and social factors that contributed to the start of the American Civil War, and compare how these factors impacted both the North and South."

There are seven key elements we can add to deepen the focus and intent of a question. Just like this example, using elements as key ingredients is essential to turning a simple starting point into a powerful tool for enquiry.

At this stage, it's critical to show students how these elements shape the depth and specificity of the questions and help them find the exact information they are looking for. Encourage them to experiment by rephrasing or stacking different combinations of elements together and seeing how each tweak shifts the intent and scope of their question.

By building their questions step by step, students see how these "ingredients" work together to craft a more compelling enquiry. In the BID routine, this practical approach to drafting questions helps students turn their initial wonderings into dynamic, purposeful enquiries. Let's examine these elements.

Temporal Elements: ⏰ ⏳ 📅

Adding a time-related element can shift a prompt from general to specific. Temporal elements allow students to look at events, concepts, or issues within a clear time frame. Whether it's about looking back or predicting forward, these give prompts context:

- **Specific Time Periods:** "During the Renaissance..."
- **Historical Periods:** "In the Victorian era..."
- **Future Predictions:** "By the year 2050..."

Comparative Elements: ⚖️ VS ☑☐

Comparative elements encourage students to explore contrasts, causes, and effects, adding an analytical layer that brings depth to their questions. These elements are ideal for helping students draw connections and notice distinctions:

144

- **Compare and Contrast:** "Compare and contrast..."
- **Cause and Effect:** "What are the causes and effects of..."
- **Similarities and Differences:** "How are X and Y similar and different?"

Spatial Elements:

With spatial elements, students add a sense of place to their prompts. It's perfect for questions where the 'where' is essential to understanding the subject:

- **Geographic Location:** "In the Amazon Rainforest..."
- **Urban vs. Rural:** "In urban and rural areas..."
- **Ecosystem Focus:** "Within coral reef environments..."

Perspective Elements:

These elements bring in different viewpoints or ethical considerations, giving prompts a more complex, multi-dimensional feel. They help students consider alternative lenses and question the bigger "why" behind issues:

- **Different Viewpoints:** "From the perspective of..."
- **Ethical Considerations:** "What are the ethical implications of..."
- **Cultural Perspectives:** "How would indigenous communities view..."

Analytical Elements: 🏛 🔍 📊

Analytical elements prompt students to dig deep, evaluating arguments, weighing evidence, and looking critically at different aspects of a topic:

- **Analyze and Evaluate:** "Analyze the strengths and weaknesses of..."
- **Critical Thinking:** "Critically evaluate the arguments for and against..."
- **Data Interpretation:** "What trends can you observe from this data set?"

Creative Elements: 💭 🎭 💬

Creative elements invite students to imagine and explore hypothetical situations, fostering out-of-the-box thinking and inspiring new ideas:

- **Hypothetical Scenarios:** "Imagine a world where..."
- **Creative Writing:** "Write a short story about..."
- **Future Predictions:** "What might the future look like if..."

Practical Elements: ✂ ⚙ ♀

Practical elements bring a hands-on problem-solving approach, perfect for prompting students to think through real-world issues and design actionable solutions:

- **Problem Solving:** "How can we solve the problem of..."
- **Decision Making:** "What is the best course of action for..."
- **Design Thinking:** "Design a solution to the problem of..."

By understanding these different prompt elements, students have a reference point to refine their queries and return better responses from AI. It's like adding layers to an onion, where each element adds depth and complexity to the prompt.

And about those emojis - yes, they serve a purpose beyond just being fun. Imagine trying to teach an entire class to remember prompt elements like "temporal" or "spatial" without a little help! Emojis act like memory aids, giving students a quick, visual "hook" to recall key elements. Plus, let's be honest: They're more engaging and recognizable at a glance than a textbook definition. Whether you're teaching younger elementary students or working with diverse learners, a simple icon can be all it takes to turn "temporal" into ⏰, or "creative" into 🐱 - and that makes these ideas easier for everyone to access.

In fact, these symbols work perfectly on anchor charts around the room, or as little desk reference cards in table groups and visual reminders at peer partner stations. They could even become part of a 'Prompt Swipe File' that students build as a personal reference resource. My confession: I hoard prompts like Carrie hoarded shoes. I'm not ashamed of that 👠, but I am sad if you missed the reference.

Emojis also make it easier to reach different age groups and accommodate various learning preferences. They give students a fun, low-pressure way to connect with these elements, especially for

148

younger learners or those with unique learning needs. Sometimes, a little icon is all you need to make abstract concepts stick.

Love these elements and their emojis? If you would like a printable copy of these, visit www.peaceframework.com/promptingelements

As teachers, we are always looking for ways to have students collaborate in an inclusive environment. When students are drafting their questions as part of their BID or ReBID routine and layering in these elements, add some extra time for pair shares or small group discussions. Peer feedback is the perfect way to help students refine further, spot blind spots, and clarify focus. So, let's shift our focus to collaboration, because the best enquiry often starts with a spark from an inspiring discussion.

Peer Collaboration

Remember that old quiz game show where contestants could phone a friend if they got stuck? I always wondered if the friend on the other end was frantically

Googling while saying, "Uh… hold on, let me think." Sometimes they'd know the answer, but other times, just talking it through helped the contestant piece things together. Collaboration works the same way in the classroom - everyone needs a sounding board.

I'll never forget one brainstorming session where a student proudly asked, "Can we interview Kevin Hart about photosynthesis?" After the excitement settled (and I tried to imagine what Kevin might say), it hit me: Sometimes the wildest questions lead to the best discussions. And that's the beauty of collaboration - it brings out different perspectives.

Embed Peer Collaboration

Why do we keep getting students to interact? Well, there is one thing we've become acutely aware of with the rise of AI: All the ways a student can 'cheat' their way out of our favorite go-to assessments. Communication and collaboration are the obvious loopholes. When our students talk through problems, collaboratively work on questioning, and then refine ways to disagree and still arrive at resolution, it helps them build all those critical skills we so desperately need to cultivate in our students without an AI power up. So, let's talk about collaboration, because everyone needs a sounding board.

And as a sidebar, calling Kevin Hart about photosynthesis might actually benefit the discussion with a bit of perspective. What if he had an interesting take on it, a story that got us thinking, or a question that took us in a different direction. You never know! But what's clear is that in a collaborative setting, we benefit from different perspectives, which can help pinpoint investigable questions.

So, how do we work collaboration into the Enquire phase? Simple: *Small groups and peer partnerships.*

Peer Partnerships & Small Groups

Let's get into the details here, starting with peer partnerships. First, students do a quick question storm to start their BID question protocols. Then, students pick some questions that stand out and explain to a partner why they think these questions have enquiry potential. They share their thought process, the rationale for selecting each question, and even predicting possible outcomes. It's an approach that encourages students to take ownership and engage in dialogue that goes beyond "is this a good question?"

Instead, it's "Here's why I think this question could go somewhere interesting."

In small groups, students can go a step further, taking turns to share their top questions and exploring each one from different angles. For example, students might ask each other, "What do you think this question might tell us?" or "How could we figure this out?" Here are a few more stems to teach your students:

- "I think this question is interesting because..."
- "This question could lead us to discover..."
- "What caught my attention about this question is..."
- "If we dig into this, we might learn..."
- "I see potential in this question because..."

By explaining their choices and hearing their peers' feedback, students build a collaborative lens on enquiry. And they're not just identifying interesting questions - they're learning to articulate their reasoning and predict where each question might lead.

You might be freaking out trying to figure out how to fit all this collaboration time into your finite instructional minutes. Don't worry, this isn't meant to be an all-day affair! This cycle of question-collaboration can be streamlined into just a few minutes and is easier to incorporate than you think.

Let's break it down using the BID method:

- **Brainstorm:** Spend 2-3 minutes to brainstorm as many questions as possible. Focus on speed and creativity, using your target focus area or a thought-provoking opener.
- **Identify:** Give students 2 minutes to share their questions with a peer or small group and identify the ones with the most potential.
- **Draft:** Use 2-3 minutes for students to draft their top question by adding elements and get ready for the next step.

In just 6-8 minutes, you've turned wide-ranging curiosity into actionable questions, all while making the most of your class time. Even better, your students are now focused on the lesson, engaged through peer collaboration, and excited to tackle the questions that matter most to them.

The key to success at this point is clear modeling and lots of practice. This is where we, as teachers, come back into the picture. What strategies can we use to be more effective models? Well, I'm glad you asked.

Model Like You Mean It

Remember learning to ride a bike? Chances are, you didn't start by reading a manual on bicycle dynamics. Someone showed you how it's done, steadied your handlebars, and ran alongside until you found your balance. The same goes for questioning in our AI-powered classrooms.

I learned this lesson the hard way during my first attempt at traditional inquiry-based learning. I'd given them clear instructions and a little spark, but the questions they asked were about as deep as a puddle in the desert. *"What is a tundra?" "List animals in the rainforest."* Do these Googles sound familiar?

That's when it hit me - I hadn't shown them how an expert questioner thinks. I was essentially asking them to ride a bike after just reading the instruction manual.

Here's where transformation happens:

Think Aloud, Think Often

Instead of just demonstrating good questions, I started narrating my questioning process: "Okay, we're studying biomes. My first instinct is to ask 'What is a tundra?' but let me think... What do I really want to

154

know? I'm curious about how things survive there. Maybe I should ask about adaptation strategies... or wait, even better, what if we compared survival strategies across different extreme environments, like the tundra versus the deepest depths of the ocean?"

Make Your Mistakes Visible

Students need to see that even teachers struggle to craft perfect questions. I'll often intentionally start with a basic question and then critique it out loud: "Hmm, 'List the animals in the rainforest' - that's where my brain went first. But would that really help us understand rainforest ecosystems? What if instead..."

Show the Power of Iteration

Demonstrate how questions evolve through multiple drafts:

- Draft 1: "What animals live in the rainforest?"
  Think aloud: "That's pretty basic. Let's focus it more..."
- Draft 2: "How do rainforest animals survive?"
  Think aloud: "Better, but still broad. What specific aspect..."

- Draft 3: "How do different layers of the rainforest canopy create unique survival challenges for animals?"

In the next section, we'll introduce the SHARP Filter, which helps systematize this process of question evaluation. But remember, this only works when students have seen questioning in action first.

Think of it this way: If questioning is like riding a bike, modeling is holding the handlebars while your students find their balance. The BID routine and the SHARP filter are like the training wheels that give them confidence to ride solo. And AI? That's like letting your students ride an electric bike. They still need to know all the fundamental skills, but now they're on something that will take them further than they could ever go before.

Ready to start modeling? Try this tomorrow: Pick one moment in your lesson to pull back the curtain on your questioning process. Let students see how your brain moves from basic to brilliant questions. Trust me... they'll be crafting deeper questions before you know it.

Classroom Vignette: BID in First Grade

In Ms. Rivera's first-grade classroom, students are discovering how to ask thoughtful questions through the BID routine, learning to wonder about everyday animals and exploring enquiry with AI-like thinking.

The Setup

A colorful "Question Wall" displays their BID process:

- **Brainstorm:** Ask ANY question!
- **Identify:** Pick our best questions
- **Draft:** Make our questions shine

Modeling   in   Action

During center time, Ms. Rivera gathers a small group to explore the topic - animals in our neighborhood. She introduces the brainstorming phase with enthusiasm, "Let's start our BID and see how many questions we can come up with! Remember, there are no silly questions!"

157

The first-graders eagerly share:

- "How do birds stay warm?"
- "Can raccoons play tag?"
- "Do squirrels wear pajamas?"
- "Where do dogs go to sleep?"
- "Why do cats have whiskers?"

Refinement Practice

As a group, students identify the question they think is most interesting. This small group picked the squirrels and pajamas. Ms. Rivera affirms that choice, "I love this one! Let's get it AI ready and draft a detailed question."

Through guided discussion, the group works together to transform the brainstormed questions they identified into a more focused and meaningful enquiry:

- Initial: "Do squirrels wear pajamas?"
- Refined: "What do squirrels do at night?"
- Final Draft: "How do squirrels make their homes and stay safe in the dark?"

Ms. Rivera's classroom doesn't have individual devices for every student nor do any of her student devices have access to AI large language models. Despite that, she models how to ask AI by acting as a digital assistant herself. "If we typed 'Do squirrels wear pajamas?' into the computer, it might say something funny... but let's think about the time element and ask "How do squirrels make their homes and stay safe in the dark?" and find out the answer!

Class Collaboration

Each small group contributes their final questions to the shared *Question Wall*, a growing anchor chart that showcases the class's collective curiosity. At the end of the unit, she will read some of those questions aloud and the students who helped write those questions get to share the answers with classmates. Later, as part of future research projects for this unit, students will choose their favorite question and ReBID to dive deeper into something that sparked their curiosity.

Teacher Reflection

"The BID routine makes enquiry accessible and engaging, even for my youngest learners. By walking them through brainstorming, refining, and drafting, they see how their questions evolve into something meaningful. It's amazing to watch their excitement as they add their best questions to our Question Wall. Even without personal devices or direct access to AI, we're building their skills to engage with technology in a thoughtful way and laying the groundwork for learning how to learn."

Making Modeling Stick: Your Game Plan

You know that feeling when you leave a great professional development session full of ideas, only to have them fizzle out by Monday? Let's make sure that doesn't happen with your question modeling practice. Here's your practical game plan:

1. **Game Plan #1: Sandbox Practice Time**
   Before rolling out your question modeling with students, take 20 minutes to practice on your own. This is your time to experiment, refine, and adjust without any pressure.
   Steps:
   - Choose a topic you'll be teaching next week.
   - Write down your initial content questions.

- Practice thinking aloud as you refine and evolve your question.
- Try a few rounds with AI to see where it helps (and where it doesn't) as you streamline your questions.

Think of this as your dress rehearsal; better to iron out the kinks now than in front of 25 eager faces!

2. **Game Plan #2: Strategic Model Moments**
   Instead of random modeling, strategically insert these moments into your weekly plan. Add places in your lesson throughout the week where you can model your question-building process as you slowly make it part of your routine.
   Steps:
   - Start by asking a surface level question that is simple and factual, like 'What is photosynthesis?" Write it on the board or on a digital screen to preserve the starting point.
   - Then walk through how you would refine the question by thinking aloud. "Let's think: Photosynthesis is about how plants use sunlight to make food. What's the

next step? How can we make this question more focused and specific?"

- Finally, identify the refined question, like "How do plants capture sunlight for photosynthesis?"

BONUS: Once you feel comfortable with this level of modeling, engage students in the process! Steps:

- Start with that surface level question and write or post it.
- Ask students to brainstorm suggestions on how to refine it, guiding them with discussion prompts like:
  - "What additional detail could we add to make this question more specific?"
  - "Instead of just asking *what*, could we think about *why* or *how*?"
  - "Is there a particular part of the process we could focus on to ask a more interesting question?"
- Encourage students to help draft aloud and share their ideas. Students might say things like:

- ○ "What about different types of plants? How do they capture sunlight differently?"
- ○ "Maybe we could add the word 'effectively.' How do plants *effectively* capture sunlight?"
- Capture this evolved, specific, and refined question on the board or digital screen to compare it to the original question and evaluate it.
- Recap with students by explaining the steps of reviewing the initial question, brainstorming ways to focus and be more specific, and then drafting a more meaningful question.

3. **Game Plan #3: Question Evolution Station**

Set up a visible space (physical or digital) where you can track and display the development of your question.

Steps:

- Label areas: Basic ➜ Better ➜ Best
- Ask your 'basic' level question, placing it in the appropriate section.
- Model with self-talk the process of adding detail and focus as you draft more

163

questions, placing them in the appropriate section.

TIP: Prepare a sample set of questions for each category, like:

- ○ Basic: What is photosynthesis?
- ○ Better: How do plants capture sunlight?
- ○ Best: How do different leaf structures optimize light capture across ecosystems?

This will give your students a tangible way to see how questions evolve and improve.

4. **Game Plan #4: Craft Your Think-Aloud Scripts**

   Keep a set of prompts on hand for your think-aloud process. These will guide you as you model how to refine questions, making your thinking visible to students.

   Prompts to Use:
   - "My first thought was [basic question], but what if we..."
   - "That's interesting, but could we make it more specific by..."

- "I'm noticing this is just a fact-finding question. How could we dig deeper?"

These scripts will help you stay focused on refining your questions while also modeling the process for students.

Your next steps: Pick **one game plan** to implement tomorrow. You don't need to do everything at once! Focus on consistency, and gradually add in the next strategy the following week. In order to effectively apply the rest of the PEACE Framework, we need to nail this modeling foundation first and work towards being better questioners ourselves. This isn't about perfection - it's about progress. The beauty of this game plan is that it grows with you and your students. Each class, each question, each iteration builds your collective capacity for deeper thinking.

You may be a bona fide superstar teacher who wants to implement all 4 of these gameplans all at once. Here is an example:

You start by writing **"What is photosynthesis?"** on the board.

"Alright, class, let's start here. This is a pretty straightforward question. It's asking for a definition. But here's the thing: In science, we don't always just want definitions. We want to dig deeper and understand the *process* behind things. Now, let's think about how we can make this question better. Hmm..."

You pause for a second, thinking aloud, giving students a chance to hear your thought process.

"My first thought was 'What is photosynthesis?' because that's what we are all *supposed* to know. It's a simple question though, right? So, then I think: Can we make this more specific? Could we ask something that focuses on how photosynthesis works? Maybe we could make it more detailed by asking how sunlight is involved."

You write **"How do plants capture sunlight for photosynthesis?"** on the board.

"Okay, that's better! Now, we're asking about how plants specifically capture sunlight for the process. But what else can we do? What's another layer we can add to really make this question even more specific?"

As you think out loud, you begin to refine further.

"That's interesting, but could we make it even more specific by considering different types of plants? Maybe some plants capture sunlight differently than others. What if we ask about the efficiency of light capture across different plants?"

You pause, then write **"How do different leaf structures optimize light capture for photosynthesis?"** on the board.

"Now, that's a really detailed question. We've gone from just asking what photosynthesis is, to exploring how plants capture sunlight, to considering different leaf structures. This question will lead us to a much deeper understanding, don't you think?"

At this point, you step back and ask your students to think with you.

"Okay, class, let's take a moment. We started with a very basic question: 'What is photosynthesis?' But now we've turned it into something much more interesting. How did we do that? What did we add to make this question more focused and specific… maybe by incorporating a **spatial element** like 'how different plants grow in different environments,' or a

comparative element such as how various leaf structures optimize light capture?"

You encourage a few students to chime in.

"What details did we add? What words or phrases helped us focus the question? Did we just keep asking 'What' or did we think about 'How' and 'Why' too?"

As students suggest ideas, you continue guiding them with questions.

"Exactly! We added specifics about how light is captured, and then we focused on the *leaf structures*. By focusing on one aspect of photosynthesis, we're already thinking more deeply. Now, how could we make this question even better? Could we add a *why* or *how* related to the environment?"

Some students offer suggestions, and you write their ideas on the board as part of a discussion.

"I love those ideas! Now, here's where we're at: We've gone from asking a simple 'What is' question, to exploring how sunlight is captured, to considering the *how* and *why* of leaf structures in different plants. All of these layers help us get a clearer picture of photosynthesis. We're thinking critically about it, and that's what we want to aim for in all our questions."

To bring it all together, you refer to the **Question Evolution Station**.

"Now, let's track how our question evolved. First, we had our basic question: 'What is photosynthesis?' That's in the **Basic** column. Then we refined it to 'How do plants capture sunlight for photosynthesis?' - that's the **Better** question. And now we're at our **Best** question: 'How do different leaf structures optimize light capture for photosynthesis?'"

You show them the **Basic ➔ Better ➔ Best** chart on the board, with each of the questions placed in the appropriate section.

"See how we moved from a general question to something very specific? Each time we asked ourselves: 'How can we make this better?' and 'What else can we focus on?' That's the process of refining questions."

To wrap it up, you offer one last think-aloud prompt to help students understand the progression.

"Remember, when I started, my first thought was just a basic question; something easy to answer. But through this process, I kept asking myself: 'Can I make this more specific? Can I focus on a particular aspect? How can I dig deeper into the topic?' And that's how you build better, more interesting questions. Now, I want you to think about your own questions. What's something you can refine today?"

If this seems like a lot, then start with the smaller game plans to model for your students how questioning evolves. You can do this! And, trust me, your future self (and your students) will thank you for laying this groundwork now.

So far, we've walked through the Enquiry process by using the BID routine (**B**rainstorm, **I**dentify, and **D**raft) to create purposeful enquiries by brainstorming ideas, identifying key concepts, and drafting questions. We used the ReBID routine to ask more specific and focused questions by adding elements, like spatial, temporal, or comparative details. We've also incorporated lots of collaboration, using peer feedback throughout the questioning process to make learning more engaging, interactive, and ultimately

170

AI-proof. Now that we've crafted and refined our questions, it's time to step back and evaluate. This is where we assess whether our question is ready for the next phase of the PEACE process or if it needs to be ReBID again and reworked before moving forward. So how can students evaluate their question to make sure it's ready to drive meaningful exploration?

Evaluating SHARP Questions

Remember playing "20 Questions" as a kid? The game wasn't really about asking twenty questions - it was about asking the *right* questions to solve the puzzle as quickly as possible. Every question mattered.

I learned this lesson the hard way working with students on a unit about climate change. With their newly minted AI access burning a hole in their pockets, they set out to be 'curious' and generated over fifty questions in their first brainstorming session. Impressive? Yes. Overwhelming? Absolutely. We had accidentally created what I now call "question paralysis". There were so many questions that we couldn't see the forest for the trees.

That's when I developed the SHARP Filter. Think of it as a mental strainer that helps catch the golden questions while letting the less effective ones flow away:

The SHARP Questions Filter

The SHARP filter lets us find the best questions through evaluation by asking:

- *Specific*: Does the question target a precise aspect of the topic?
- *High-level*: Does it promote higher-order thinking?
- *Actionable*: Can the question be researched with available resources or evidence?
- *Relevant*: Does it connect to our learning objectives and real-world applications?
- *Powerful*: Does it have the potential to generate new questions and deeper understanding?

Here's how it works in practice.

Progressive Example

Let's say your class is studying climate change. After the Provoke phase in PEACE, they've brainstormed tons of questions and are wondering which ones to follow up on. We can run their questions through the

SHARP filter and eliminate the unsharp ones. Here are some of the questions:

- What is climate change?
- What are the causes of climate change?
- How does climate change affect ecosystems?
- What are the potential solutions to climate change, and which ones are most effective?
- How can we implement sustainable practices in our local community to mitigate the effects of climate change?

Now, run them through SHARP one at a time:

Question 1: What is climate change?

- **Specific:** Not very specific, too broad.

This question is eliminated.

Question 2: What are the causes of climate change?

- **Specific:** More specific than the first question.
- **High-level:** Doesn't encourage deep thinking, focuses on facts.

This question is eliminated.

Question 3: How does climate change affect ecosystems?

- **Specific:** More specific than the previous questions.
- **High-level:** Encourages deeper thinking, requires analysis.
- **Actionable:** Because it is so broad in scope, research will likely be overwhelming.

This question is eliminated.

Question 4: What are the potential solutions to climate change, and which ones are most effective?

- **Specific:** More specific than the previous questions.
- **High-level:** Encourages critical thinking and evaluation.
- **Actionable:** Researchable with accessible evidence and resources.
- **Relevant:** Relevant, but lacks a specific focus.

This question is eliminated.

Question 5: How can we implement sustainable practices in our local community to mitigate the effects of climate change?

- **Specific:** Very specific, focuses on local action.
- **High-level:** Encourages critical thinking and problem solving.
- **Actionable:** Researchable with a focus on local resources and practical evidence.
- **Relevant:** Highly relevant to local communities.
- **Powerful:** Now that we have a question which cleared the first four checks, we can ask ourselves if it sparks our curiosity and seems worthy of further enquiry.

If the answer is 'yes', we've just managed to cull our long laundry list of questions into a focused enquiry about the topic that provoked us.

By applying the SHARP filter progressively, students learn to filter out basic surface-level questions and focus on the most powerful questions that can drive meaningful learning and positive change. But what if you're thinking, 'I'm not sold on a progressive filter.' That's okay! We can approach this as a checklist-style evaluation.

Checklist Example

If you're wondering if you can use SHARP filters to find the *sparkliest* questions (I mean, who wouldn't want the shiniest ones, right?) The answer is YES! We can absolutely put each question through the wringer with a checklist approach to see if it's got the shine. To make it easier to evaluate, let's turn each criterion into a simple yes or no, allowing us to score a question with a maximum possible score of 5.

Let's try this checklist and scoring approach with questions on ocean pollution:

Question 1: What is ocean pollution?

- **Specific:** No. While it's a good starting point, it's too broad.
- **High-level:** No. Doesn't encourage deep thinking.
- **Actionable:** No. Broad scope doesn't make it immediately actionable.
- **Relevant:** Yes. Relevant, but lacks specificity.
- **Powerful:** No. Could spark further questions, but it's not particularly powerful.

This question only scored a 1/5, meaning it needs significant work. After going through the checklist, it's clear it's not SHARP and a dud for our PEACE plan. But here's the thing about question development... sometimes our "okay" questions are just stepping stones to great ones. Let's try another:

Question 2: What are the effects of ocean pollution on marine life?

- **Specific:** No. Somewhat specific, but could focus even more.
- **High-level:** Yes. Encourages deeper thinking.
- **Actionable:** Yes. Researchable with evidence and data.
- **Relevant:** Yes. Relevant, but could add more specificity.
- **Powerful:** No. May have the potential to spark further enquiry and discussion.

We can see a shift towards SHARPening the original question. This one gets a 3/5, which is better. Now we have to ask ourselves if we are done or if we want to SHARPen the question more and get to that perfect five out of five.

Question 3: How do different types of ocean pollutants interact with marine ecosystems, and

which poses the greatest threat to our local coastline?

- **Specific:** Yes. Very specific with clear focus.
- **High-level:** Yes. Encourages deeper thinking.
- **Actionable:** Yes. Researchable with evidence and data.
- **Relevant:** Yes. Relevant to our community.
- **Powerful:** Yes. Has the potential to spark further enquiry and discussion.

Now we're getting somewhere! This question is SHARP and with a 5/5, has serious potential! When students learn to evaluate their questions using these filters, they're crafting strong enquiry and also developing a critical thinking mindset. Using the SHARP filter as a checklist gives them a consistent way to evaluate their questions and transform them into curiosity-driven enquiry that challenges their initial assumptions and drives deep learning.

Classroom Vignette: SHARP Question Refinement

In Ms. Patel's 8th-grade social studies class, students are learning to transform basic questions into powerful enquiries about the Industrial Revolution.

The Setup

The **SHARP** criteria are displayed prominently on the classroom wall:

- **Specific**: Targets precise aspects of the topic.
- **High-level**: Promotes deeper thinking and analysis.
- **Actionable**: Can be researched and answered.
- **Relevant**: Connects to key objectives and real-world connections.
- **Powerful**: Generates new questions and open up further enquiry.

Ms. Patel is eager to help students refine their questioning skills, ensuring they can ask questions that will lead to deeper engagement and discovery.

Modeling in Action

Initial Student Question: "Why did factories change things?"

Ms. Patel reads the question aloud, thinking through each SHARP filter:

- **Specific**: "Not really. 'Things' is too vague. We need to narrow it down to something more precise."
- **High-level**: "It's a good starting point, but let's see how we can push it to promote deeper thinking."
- **Actionable**: "Can this question be explored effectively with the resources we have?"
- **Relevant**: "This question is relevant, but we need it to tie into the larger themes of the Industrial Revolution."
- **Powerful**: "Can this question spark new lines of enquiry? I think we can make it more powerful."

Ms. Patel types into Claude, her favorite LLM: *"Help me transform this question about Industrial Revolution factories into one that meets SHARP criteria."*

After Claude returns a response, the students evaluate it against the SHARP filter and like their new question:

"How did the introduction of factory systems in the 1800s reshape both working conditions and family structures in urban communities?"

Ms. Patel explains: "Notice how this question is more specific? It's focusing on two key aspects - working conditions and family structures. Working conditions is a temporal element that examines change over time and family structures is a comparative element to compare and contrast. This question is high-level because it encourages us to think critically about change. It's actionable, since it leads to clear areas of investigation, and it's powerful because it opens up lots of ways to question further."

AI-Supported Practice

Now, it's the students' turn to practice using AI as a questioning partner. Ms. Patel gives each pair of students the following steps:

- **Step 1**: Input their initial questions into the AI tool.
- **Step 2**: Review the SHARP-aligned suggestions the AI provides.

- **Step 3**: Evaluate and modify the AI-generated questions to make them their own with an element.
- **Step 4**: Share their refined questions with a peer partner or small group for feedback and further improvement.

Sample Student Progress: Jayden's Question Evolution

Starting: "What did workers do?"

AI Suggestion: "How did worker responsibilities change?"

Add Elements: Add a temporal element for time-specific context and comparative elements with specific jobs and focused impact.

Final Refined: "How did the shift from skilled craftwork to specialized factory roles impact worker autonomy and job satisfaction in the 1840s?"

Jayden's Discussion Board Post:

"At first, I thought using SHARP questions was just going to be extra work. Like, why can't we just ask normal questions? But after working with AI, I started to see how better questions lead to way better answers.

When I changed my question from 'What did workers do?' to asking about how factory jobs changed people's lives, there were so many interesting stories about real families. Now, my dad says I ask too many questions at home 😆, but that's what historians do, right?"

Teacher Reflection

"Using AI as a questioning partner gave students immediate feedback while they practiced. But I think the real power came from making the thinking process visible. They saw how basic questions could evolve into more challenging and meaningful ones. Plus, they made connections between what they were learning and their own experiences. And the best part? The technology didn't replace critical thinking... It enhanced it! AI was just the tech tool; they were the ones in charge of their own learning."

Special Education Spotlight: Making this process work for every student matters. For our students who may struggle with this process, try creating a visual SHARP card with icons representing each filter element. I've seen amazing results when students

183

physically move their question cards through each checkpoint and quickly reference the emoji… it makes the abstract concrete and manageable.

Here's a fun way to attach visuals to this process, along with some simple self-check questions that students can ask themselves:

- **Specific:** ◎ (bullseye)
  - ⚮ "Is my question focused on one clear thing?"
  - ☞ "Could I point to or picture exactly what I'm asking about?"
  - Think: 📷 "Am I using a zoom lens or a wide-angle lens?"
- **High-Level:** 🧠 (brain)
  - 🤔"Does my question make me really think, or is it just about facts?"
  - 🔵"Can I Google the answer in 5 seconds?"
  - Think: 💬"Would this question start an interesting conversation?"
- **Actionable:** 🚀 (rocket)
  - 🔍"Can I actually find or figure out the answer?"

- o 📖"Do I know where to start looking?"
- o Think: 🔍"Is this something I can actually investigate?"
- **Relevant:** 🌐 (globe)
  - o 💚"Does this question matter to me or others?"
  - o 🔗"Can I connect this to something in real life?"
  - o Think: 👍"Would someone else care about the answer?"
- **Powerful:** 🔥 (fire)
  - o ♾️"Does this question lead to more questions?"
  - o 🔄"Could the answer change how I see things?"
  - o Think: 😮"Will the answer make me say 'wow'?"

Here's a pro tip: If you're working with students who are easily overwhelmed, simplify the visuals by just working towards a quality self-reflection with an emoji anchor:

When you think you have a great question, ask yourself:

- 👀 Is it clear and focused? It's **S**pecific.
- 🙂 Will it make me think deeply? It's **H**igh-level.
- 🔍 Can I actually find an answer? It's **A**ctionable.
- 💜 Does it matter to me or others? It's **R**elevant.
- 😮 Might the answer make me say 'wow'? It's **P**owerful.

If you can say "yes" to most of these, you've got a SHARP question!

I know all of this can feel overwhelming or out of reach, but everything in this book is about progress over perfection. If you throw this whole SHARP filter with a million emojis at your students like confetti on New Year's Eve, all you're going to get are the wide eyes of thirty deer in headlights. Move slowly through the PEACE Framework, building skills within each element, by having authentic ruler-worthy wonder moments, asking questions through the BID routine, adding elements within the ReBID process, and finally applying a SHARP filter to find the questions worthy of enquiry.

SHARP Implementation Guide

Start by focusing on one SHARP filter at a time to avoid overwhelming students. The goal is to make the process manageable and help students build their skills incrementally. Most teachers begin with the Specific filter and introduce each subsequent filter gradually, spending 1-2 weeks on each one.

Phased Implementation Strategy

- **Week 1-2:** Focus on the **Specific** filter
- **Week 3-4:** Introduce the **High-Level** filter
- **Week 5-6:** Move on to the **Actionable** filter
- **Week 7-8:** Teach the **Relevant** filter
- **Week 9-10:** Wrap up with the **Powerful** filter

This progression allows students to practice and refine their questioning without feeling overwhelmed. Remember, not every lesson needs a SHARP question. Sometimes practicing one filter in depth is a huge win.

1. Specific Filter

The first step is to teach students to recognize broad versus focused questions. Try practicing the process of narrowing the scope before adding complexity.

Sentence Stems:

- "Can you zoom in on exactly what you want to know?"
- "What specific part of this are you curious about?"
- "Narrow this down to one clear thing."

Teacher Support: Model narrowing broad questions. **Example:** "Eating" → "Eating during flood season"

2. High-Level Filter

Encourage students to move beyond simple fact-gathering by asking themselves how the question could lead to deeper thinking. Help them make connections to bigger concepts.

Sentence Stems:

- "This could be deeper. How can you make this more thought-provoking?"

- "What bigger idea is behind this question?"
- "How might this connect to larger concepts?"

Teacher Support: Push students beyond just fact-gathering.

Example: "What do farmers eat?" → "How do eating habits reflect social structures?"

3. Actionable Filter

Ensure the question can realistically be explored or researched. Does it invite investigation and discovery using available resources?

Sentence Stems:

- "Can this question be explored effectively with what we have?"
- "What tools or sources could help you investigate this further?"
- "Is this question clear enough to guide your research?

Teacher Support: Help students think about possible research pathways.

Example: Provide mini-lessons on how to prompt AI and use Google searches.

4. Relevant Filter

Encourage students to think about why their question matters. How does it connect to their lives, the world around them, or current events?

Sentence Stems:

- "Why does this matter to me/others?"
- "How does this connect to real-world experiences?"
- "What makes this question meaningful?"

Teacher Support: Link questions to student interests and current events.
Example: Relate questions to ongoing discussions or personal interests.

5. Powerful Filter

Help students recognize that great questions often lead to more questions. Encourage them to think about how the answer could change their perspective or lead to unexpected insights.

Sentence Stems:

- "Could this question lead to more questions?"

- "Might this change how you understand something?"
- "What surprising insights can you uncover?"

Teacher Support: Encourage speculation and curiosity.

Example: Use open-ended prompts to push students toward deeper thinking.

Practical Classroom Techniques

Want to get your students learning about applying the SHARP filter? Try these:

- **Quick Question Refinement Mini-Lessons** Teach students how to quickly refine questions during short, focused sessions. Use the sentence stems to guide them through the process.
- **Question Improvement Stations** Set up stations where students work in groups to refine questions using the SHARP filters. Rotate through stations to get multiple perspectives.
- **Collaborative Question-Crafting Activities** Have students work together to craft SHARP questions about a topic. Provide opportunities for peer feedback to encourage collaboration and critical thinking.

- **Question   Evaluation   Station**
  Set up a station where students read through a set of questions and apply the SHARP filter as a checklist to score them on a scale of 1-5. Building a SHARPer eye can also spark curiosity and fill prior knowledge gaps.

The goal is progress, not perfection. Don't expect students to hit all five filters at once. Celebrate the small wins... like successfully narrowing a broad question or making it more thought-provoking. By introducing the filters one at a time, using sentence stems for support, and providing plenty of opportunities for practice and feedback, students will refine their questioning skills over time and develop deeper, more meaningful enquiries.

An AI-Powered Tip

Remember that math teacher who insisted you show all your work? The one who'd write "Show your steps!" in red pen while you rolled your eyes because hey, calculators exist! Well, plot twist... I'm about to channel that teacher's energy, but with a modern upgrade.

Up until now, we've been working through SHARP filters the old-school way. And yes, that's important! Just like understanding long division helps you spot check your calculator, understanding how to manually SHARPen questions helps you evaluate AI-generated ones. But here's where it gets fun. We can actually turn AI into our question-refining sidekick.

Here's a prompt you can give your students:

Prompt: "I have this basic question about (topic): (insert questions). Help me apply these criteria (insert SHARP filters) to develop three better questions."

Check out the results:

Prompt: "I have this basic question about ancient Egypt: "How did they build the pyramids?" Please help me develop this into three stronger questions using these criteria:
- Specific: Focus on particular aspects or time periods
- High-Level: Require analysis and deeper thinking
- Actionable: Can be researched or investigated
- Relevant: Connect to broader themes or modern life
- Powerful: Lead to meaningful insights
For each suggested question, explain which SHARP criteria it meets and why."

And boom! The AI helped us transform that basic question into gems like:

1. "How did innovations in ancient Egyptian mathematics and engineering during the Old Kingdom enable the progression from step pyramids to true pyramids?" (**S**pecific to time period and technology, **H**igh-level analysis, **A**ctionable through archaeological evidence.)

2. "What does the evolution of pyramid construction techniques reveal about the development of ancient Egyptian social organization and project management?" (**H**igh-level thinking, **R**elevant to modern project management, **P**owerful insights into societal development.)

3. "How do the problem-solving approaches used in pyramid construction compare to modern engineering challenges in extreme environments?" (**R**elevant to current issues, **P**owerful applications, **A**ctionable through comparative analysis.)

This is important: The goal isn't to eliminate all "basic" questions… They're necessary stepping stones, and we can use them as part of our process. The SHARP goal is to help students recognize and develop questions that will lead to deeper understanding and more meaningful enquiry. And now that AI has generated some great questions, we want students to pause. *We aren't just going to take what the AI gives us, are we?* No. Let students rate these new questions. Do they really meet all the SHARP criteria? How could we make them even better?

Key Classroom Actions:

- Introduce SHARP gradually, one filter at a time
- Model the filtering process regularly
- Create visual aids for each SHARP component
- Use AI to help teach question evolution
- Celebrate when students identify exceptionally SHARP questions

The beauty of the SHARP filter system is that it teaches students to be their own question editors. Over time, they'll naturally start generating SHARPer questions from the get-go, saving valuable class time and leading to more focused, productive enquiry.

In a Nutshell: The Enquire Phase

Remember when you were a kid and your favorite word was "why"? That's the energy we're channeling in the Enquire phase, but with a bit more sophistication and a lot more strategy. Here's how it unfolds: We start by teaching students the art of question crafting, guide them through structured BID protocols with a ReBID that includes elements, model our own questioning process, and then use the SHARP filter to refine their enquiries. Think of it as taking their natural curiosity and giving it professional training. All the while, we're incorporating collaboration and peer partnerships to use conversation as a way to further enquiry.

Here's what it looks like in action:

"Did you know there are fish that live in boiling hot water? How do they survive in conditions that would cook us alive? Think about animal adaptations and questions you might have."

"Now, let's make our opening BID. Take three minutes to write down every question you have about animal adaptations - no judging, just asking!"

"Meet with your think time team and refocus your questions by sharing your ideas, getting feedback, and adding elements to ReBID your questions as a team."

"Back together now. Watch how I add elements to this basic question in order to improve it: First I add a temporal element to 'Do dolphins sleep?' making it … 'How have dolphins' sleep patterns evolved over time?' That's good, but how about adding a comparative element to turn this question into 'How have dolphins evolved to balance their need for rest with their need to surface for air?'"

"Time to put your questions through the SHARP filter. Does your question make you think deeply? Can you actually investigate it? Will the answer matter? If they aren't SHARP, ReBID to refocus and share with your team."

Remember those rabbit holes we talked about earlier? The Enquire phase isn't about avoiding them entirely… It's about making sure we're diving into the right ones. Here's how all the pieces fit together:

- **Protocol Power:** Using structured routines, like BID (Brainstorm, Identify, Draft), isn't about putting creativity in a box - it's about providing a scaffold for students to build upon their ideas and transform them into focused, insightful questions.
- **Peer Collab:** By sharing ideas and receiving feedback, students can refine their questions, hear

different perspectives, and develop a deeper understanding.

- **Elemental Iteration**: This is where we transform "Google-able" questions into thought-provoking enquiries using the ReBID routine to add elements to our question.

- **Model the Process**: When we think aloud about our own questioning journey, we're showing students that even expert questioners iterate and refine.

- **SHARP Thinking**: This isn't just another acronym to memorize - it's your students' quality control system for powerful questions.

When you see students starting to challenge their own questions before you do - that's when you know the Enquire phase is working its magic. They're not just asking questions anymore - they're crafting them with the precision of an artisan.

Remember, there is no 'right' answer here, just 'right' questions.

Next up: *The Analyze phase, where we use these questions to uncover reliable information and become knowledge builders.*

***

Analyze

Has a student ever asked you a "simple" question about why leaves change color, only to end up on a tangent about photosynthesis, seasonal changes, and cell biology? Yeah, no. Never. But that's because we've stopped wondering in the classroom since we're too busy studying for the test. This is why the PEACE Framework is not just a conversation about AI in education, it's a shift in instruction that pulls questioning and answering to the forefront.

That's exactly what the Analyze phase is about; except now, we're equipped with AI as our research and thought partner. Our students have crafted their questions with surgical precision, having worked through the Enquiry phase, which means it's time for the investigative real work to begin.

"Investigative work?", you might be thinking. "Aren't they just going to plop their question into AI and get their answer?" In the traditional inquiry-based learning system, this process of researching was undoubtedly the most time consuming. Now, with AI, we are in an ask-and-answered era. But it's not enough to simply get an answer from an AI assistant; those answers should be critically analyzed, cross-referenced with other sources, and used as a starting point for deeper enquiry.

There are two key guiding questions throughout the Analyze phase:

Question 1: *How can I evaluate the reliability of this information?* We want our students to think critically about AI-generated results. In truth, we want them to consider reliability with all the information they encounter online, regardless of the source. By applying a simple evaluation check, students learn when to trust, when to verify, and when to dig deeper.

Question 2: *How can I refine my search to get closer to what I want to know?* Learning with AI requires an iterative approach that emphasizes the value of refining search strategies based on ongoing evaluation. We want students to see research as a dynamic process of rephrasing questions, seeking alternative sources, and adapting enquiry to get closer to the intended learning goals.

Think of this phase as teaching students to become information detectives; complete with AI-powered magnifying glasses. They're not just gathering facts;

they're analyzing, questioning, and synthesizing information to become more informed and discerning knowledge-builders.

From Information Detectives to Knowledge Builders

Growing up, I remember how big a shift it was to move into the information age. The idea that all the world's knowledge was at our fingertips required everyone to rethink how we approached learning. Now, we're moving beyond the information age into the intelligence age. It used to be enough to be an information detective, searching through millions of results on Google to find answers. But those days are over. Today, we want students to build knowledge in a way that connects to their prior understanding of the world and sparks curiosity, leading to innovation.

Up until now, the PEACE Framework has focused on sparking interest, igniting wonder, and helping students craft thoughtful questions. At this stage, a student can easily drop one of those carefully crafted questions into an AI model and get an answer. It's tempting to stop there and say, "Tada! Finished!" But in reality, this is just the beginning. Once the answer to a question has been generated, it's time to take

201

things further. That begins with evaluating the results using the SAFE check.

SAFE Check

If you survived the era when Wikipedia was the enemy of every teacher (some of us are still recovering from those citation battles), then you know that when it comes to the internet, the mantra is: *"Don't believe everything you read."* We're no longer just teaching students to be information detectives; we're helping them build robust critical-thinking skills. But how exactly do we turn our students into digital truth-seekers? We ask a simple question:

Is It SAFE?

- **S: Source:** Is the source reliable and credible?
- **A: Accuracy:** Is the information correct and free from errors?
- **F: Fairness:** Is the information unbiased and objective?
- **E: Evidence:** Is the information supported by evidence?

This is an easy-to-remember check and can be adapted for students at different age levels and experience levels. Let's break down each one.

(S) Source: Is the source reliable and credible?

Even though AI is relatively young, it was adopted at a lightning pace! Self-proclaimed coders rushed to build platforms using large language models (LLMs) like ChatGPT, but many of them relied on outdated or incomplete data to quickly launch their products. Yikes! Now, we must ask: *Which AI model are we engaging with, and where is it getting its information?* Some models offer sources or citations, some link to further questions, while others may provide outdated or vague answers.

Have an honest conversation with students about which AI models you find reliable and why, and help them recognize the potential pitfalls of working with less credible ones. Here are three classroom strategies to help students evaluate LLMs:

√ Classroom Strategy: OK-AI List

How It Works: Create a classroom list where students rank different LLM models based on performance. Discuss which bots work best for different tasks, noting strengths and weaknesses.

The Goal: Have students view LLMs as products, choosing models based on their quality and suitability for specific tasks.

Note: Make it a regular task to evaluate AI models' performance. Ranking and reevaluating based on subject needs (e.g., math vs. writing) helps students learn to select the best tool for their purpose.

✓ Classroom Strategy: AI Showdown

How It Works: Have students put the same question into multiple LLMs to compare responses. Then, guide them to analyze how each model responds, looking for differences in quality, accuracy, and depth of the information provided.

The Goal: Encourage students to critically examine varying outputs and identify which model provides the most reliable and comprehensive response.

Note: This strategy helps students realize that multiple models may yield different answers and reinforces the importance of verifying information across sources.

√ Classroom Strategy: AI Fact-Check Squad

How It Works: Assign different LLMs to each student in a group. Have them prompt their AI model, then compare the results as a group. Discuss which model provided the best, most reliable answers and which ones were derivative or less effective.

The Goal: Foster collaborative evaluation skills and help students practice critical comparison between different AI sources.

Note: This strategy encourages students to actively question and verify information, strengthening their ability to evaluate multiple sources.

Ultimately, we want our students to learn the importance of staying skeptical about the source of any information and the potential motives behind the answers provided.

Guide students to ask themselves: *Who or what is the source of this information? Is it reliable, up-to-date, and trustworthy?*

(A) Accuracy: Is the information correct and free from errors?

Sometimes AI is full of it. It's not uncommon for models to "hallucinate" or generate information that sounds convincing but is completely made up or inaccurate. It's important for students to learn to question outputs and look for ways to validate results. Encourage them to consider whether the information directly answers their prompt, whether it's relevant, up-to-date, and whether it aligns with trustworthy sources. Here are two classroom strategies to try:

✓ Classroom Strategy: AI Detective

How It Works: Share AI-generated content on a topic, then have students verify the information using reliable sources. Students should check for factual consistency and validate key details.

The Goal: Teach students to fact-check and evaluate AI content by cross-referencing with trusted sources.

Note: This strategy reinforces the importance of validating information from the AI with credible, authoritative sources.

✓ Classroom Strategy: Truth-O-Meter

How It Works: Divide students into "fact-checking teams" with AI-generated content. Have them rate the accuracy of statements on a scale from "Absolutely True" to "Nice Try, AI." before they have a chance to verify. Then have them find the truth. Afterward, students compare their findings and discuss.

The Goal: Students will practice evaluating the validity of AI content and have fun while doing it.

Note: Throw in a few deliberately wrong answers to keep them sharp and encourage them to think critically about AI-generated content.

AI models, while powerful, can sometimes confidently deliver false or outdated facts. This is where the accuracy check comes in. By incorporating these strategies, students can become more discerning users of AI, actively questioning the validity of what's presented to them. The goal is for students to understand that AI is a tool and to always ask whether the output makes sense and stands up to scrutiny.

Guide students to ask themselves: "Is the data presented by the AI current, and does it accurately address the prompt? If not, what's missing?"

(F) Fairness: Is the information unbiased and objective?

AI doesn't just report. Being heavily influenced by the data it was trained on, it can reflect biases or deliver one-sided perspectives. Teach students to look for red flags, like skewed opinions, missing details, or overly simplistic explanations that suggest a lack of objectivity.

✓ Classroom Strategy: Missing Pieces Protocol

How It Works: Students analyze AI content in three passes: 1) Circle anything biased, 2) Underline logic gaps, 3) Add questions about missing info.

The Goal: Have students critically examine AI content to uncover hidden biases or flaws.

Note: Guide students to ask follow-up questions that prompt the AI for clarification or alternative perspectives, teaching them how to dig deeper.

✓ Classroom Strategy: Spot the Spin

How It Works: Provide students with two AI-generated responses to the same prompt. If the responses don't naturally include bias, stereotypes, or misconceptions, adjust the prompt to elicit such elements intentionally.

The Goal: Have students identify any biases, inconsistencies, or oversimplifications in the AI outputs.

Note: Highlight how subtle wording choices can influence the perceived tone or bias in responses, but do so in a respectful way. You don't want to be the one introducing baseless bias into your classroom.

√ Classroom Strategy: Double Perspective Debate

How It Works: Present students with an AI-generated response on a controversial topic. Divide the class into two groups: one supporting the response and the other challenging it.

The Goal: Encourage students to explore how different viewpoints might be missing or misrepresented in the AI's output, sharpening their ability to detect bias and omissions.

Note: Use this as a chance to explore how prompt phrasing might affect the balance of AI outputs.

Fairness is essential when evaluating AI outputs. By teaching students to recognize bias, identify gaps in logic, and consider alternative perspectives, we set them up to engage critically with AI-generated content and develop the skills needed to navigate a world where information isn't always objective or complete.

Guide students to ask themselves: "Does the output provide a balanced view on this issue, or is it leaning toward a particular opinion? Are there areas that seem unclear or incomplete?"

(E) Evidence: Is the information supported by evidence from a reliable source?

AI outputs should not be taken at face value. Reinforce the importance of cross-checking AI-generated content against trusted sources like textbooks, peer-reviewed articles, or expert opinions. By comparing multiple sources, students can assess the accuracy and reliability of information. Here are few strategies to help with this:

✓ Classroom Strategy: Source Showdown

How It Works: Create three stations - one with AI-generated information, one with a scholarly source, and one with a popular media article. Students rotate through stations, completing a comparison matrix.

The Goal: Have students try to identify which source was which. To increase the rigor, have students write a synthesis paragraph identifying the strengths and limitations of each source.

Note: A comparison matrix is helpful here. Try this one:

Source	Accurate Info	Easy to Understand	Biased?	Up-to-Date
Station 1				
Station 2				
Station 3				

Which station information is: AI _____, Textbook _____, and Popular Media _____?

✓ Classroom Strategy: Triangle of Truth

How It Works: We teach students to verify information using the triangle method:

- AI Output
- Traditional Source

- Expert Opinion/Primary Source

The Goal: Have students corroborate information and collect data from multiple sources.

Note: Provide a "Synthesis Sheet" like the one below to help students organize and compare information systematically.

Topic: _____

AI Says: _____

Source 1 Says: _____

Source 2 Says: _____

My Conclusion: _____

Confidence Level: _____

We're *not* trying to turn students into AI skeptics... we're developing informed AI users who know when to trust, when to verify, and when to dig deeper. It's like teaching them to be friendly but not gullible with their new AI study buddy.

Guide students to ask themselves: "Can I find a credible secondary source on this topic. How does it compare with what the AI provided? What are the similarities and differences?"

RECAP: Is It SAFE?

By evaluating the **Source**, **Accuracy**, **Fairness**, and **Evidence** of AI-generated responses, students go beyond simple fact-checking. They develop the critical thinking skills necessary to navigate, analyze, and question information in a world increasingly influenced by AI. When they check if it's SAFE, it encourages healthy skepticism, thoughtful evaluation, and a keen eye for detail so students can become responsible consumers of information in any context.

Now that our students know how to evaluate the reliability of information, it's time to help them refine their questions based on the AI-generated results so they can actually get closer to what they want to know.

Interested in having me out to your district, region, or conference to train on AI or deliver the keynote? Reach out to me at www.peaceframework.com/speak

---

CLEAR Check to Reprompt

The first time I coached my mom through using AI, she huffed in frustration after getting a response that missed the mark completely. "This is useless," she bemoaned, ready to shut down the entire conversation.

I leaned in... "Not so fast. Working with AI is exactly like playing tennis. Your first serve might not ace the return; you've got to adjust your stance, refine your technique, and keep rallying until you land exactly where you want."

Ever tried asking a question and gotten that blank stare... from a computer?

Most people give up after the first mediocre response. But AI is not a one-and-done tool. It's about having a conversation with the chatbot... not accepting the first answer, but refining your interactions to get just what you need.

This process is just as much about critical thinking and problem solving as it is about communication and writing skills. My 70-something mom? She's now better at prompting AI than my twenty-something son. How? She asks herself: 'Is it CLEAR?' This simple check (explained below) transforms her interaction from frustrating to fantastic.

A CLEAR Check for Your AI Query

It's easy to see how students, like anyone else, might type their enquiry question into an AI tool, receive a response that's incomplete or confusing, and immediately feel stuck. Frustration builds, and they often make one of two choices - abandon the question altogether or accept the AI's limited answer as correct without further exploration. Neither approach leads to deep learning or meaningful insights.

In the **Analyze** phase, we're not just using the SAFE check to evaluate the AI results… We're also teaching students to refine their queries based on the output. This reframes the process of working with AI as a dynamic conversation, rather than a one-off query.

This is where checking if your query was **CLEAR** comes into play. By following this structured approach, students can turn frustration into clarity, learning how to iterate and improve their prompts to get better results. Let's break it down.

(C) Complete: Did the AI output completely answer all the parts of the initial question?

Sometimes, the AI's response seems spot-on at first glance, but it's essential to ensure it fully addresses every part of your query. Review the output and compare it to your original question. Are all aspects covered, or did the AI miss something important?

Ask Yourself (Key Decision Point): Did the AI completely answer the question?

- **If YES**: You're good to go! Use the response confidently, knowing it aligns with your intent. Move forward to the next phase of the PEACE Framework.
- **If NO**: It's time to dig deeper. Identify what's missing, and consider how to guide the AI toward a more complete answer.

(L) Logical: Is the AI output giving logical information that relates directly to the question's intent?

A good answer doesn't just look right; it needs to make sense. Analyze whether the AI's response is logical and directly tied to the intent of your question.

Does the reasoning align with the topic, or does something seem off?

Ask yourself:

- Does this answer logically connect to my question?
- Are there any gaps in reasoning or contradictions in the output?
- Did the AI misunderstand my intent?

If the answer doesn't feel logical, you might need to ask clarifying questions or provide more context to steer the AI in the right direction.

Even with thoughtful preparation using the BID routine, incorporating elements such as temporal, spatial, or perspective elements, and then put it through the SHARP filter, it's possible your question lacked context or relied on assumptions that didn't hold true. Before abandoning the question and starting the rewrite process, consider the next step in the CLEAR check.

(E) Explain: Do I need the AI to explain or defend its output?

It's impossible to continue our conversation with AI unless we know why it answered the way it did. Sometimes, understanding the AI's reasoning helps you refine the question or identify gaps in its response. Asking the AI to explain itself is a valuable step in the process. This not only helps you understand its logic but also gives insight into how to reword your query for better results.

Ask the AI to:

- "Explain why you answered this way."
- "Can you walk me through your reasoning step by step?"
- "What evidence supports your answer?"
- "Did you address all parts of the question?"
- "Explain this to me like I'm in 1st grade."

Understanding the AI's thought process helps identify areas for improvement and refine the conversation accordingly. With this information, we move on to the last two steps of the CLEAR check, where we enhance the response by focusing on format and phrasing.

(A) Adjust: Could the AI's output be presented in a more useful format?

The information might be accurate, but its format may not meet your needs. Adjusting the output allows you to transform the AI's response into something more practical or user-friendly. Whether you need a table, a calendar, an image, a poem, or even a script, asking for the answer in a different format can make the information much more actionable.

Ask yourself:

- Would this information be easier to use in a different format?
- Could a visual representation (e.g., table, chart, calendar) make this clearer?
- Would creative formats (e.g., poem, story, script) enhance the output's usefulness?

By tweaking the output, which we will cover more in a moment, you can turn a good answer into a great one that meets your specific needs.

(R) Reword: How can I rephrase the question to get a better response?

If the output still doesn't hit the mark, it may be time to reword the question. Rephrasing helps clarify your intent, incorporate missing details, or simplify the query for better understanding. Think of this step as resetting the conversation with a more targeted focus.

Ask yourself:

- Can I break the question into smaller parts to improve clarity?
- Can I add specific details or context to guide the AI?
- Can I rephrase vague or open-ended elements of the query to make it more precise?

Rewording isn't starting over. It's about becoming a better communicator and continuing the conversation in a way that gets you the results you want. The first query is rarely perfect, and that's totally okay. It just means continuing the conversation with a better-worded query so the AI can generate an answer that fully addresses the question's intent.

Time to get strategic.

Now that we've worked through the CLEAR check, we've had time to reflect on whether the AI's response truly addressed the intent of our question. We've also reviewed the parts of the question itself and examined why the AI answered the way it did. This is where we take all of that information, step back, and talk through it with our AI.

Here comes the moment of truth: How do you reword the question to get better results?

- **Be more specific**: Even though we used our SHARP filter to find the best question, once we start interacting with AI, we might realize that we weren't specific enough. Now that we understand AI's approach better, can we add more detail to clarify our intent?
  Prompt: *What additional information can I give you to get a better response?*
- **Chunk the question**: We added elements to our question to make it more detailed, but sometimes that extra detail can confuse the AI or result in it focusing on just one aspect. In this case, try breaking the question into smaller parts. Tackle one element at a time and then combine the results later for a more complete picture.

Prompt: *Let's break this down. Answer the first part of the question first, then we can do the rest.*

- **Change perspectives**: Sometimes, the AI may have approached your question from a different angle than you intended - perhaps it responded like a scientist when you wanted a policymaker's perspective. By switching the perspective, you can see if you get a more relevant or insightful response.

Prompt: *Think of this as a [perspective] and not a [perspective].*

- **Clarify the scope**: If the answer feels too broad or too narrow, the AI might need a clearer sense of the scope. Are you asking for general information, or do you need a deep dive into a specific topic? Adjust your question accordingly to reflect the level of detail you need.

Prompt: *Can you focus more on [specific aspect]? Or could you give me a more general overview?*

- **Refine the wording**: A small change in wording can make a huge difference in the AI's response. If the initial phrasing of the question didn't quite work, rephrase it using simpler or more direct language. This can help the AI better understand your query.

Prompt: Let's reword this question a bit. Can you focus on [key term] rather than [original term]?

- **Change the output:** We are going to cover this more in the next section, but it is possible that the response wasn't helpful because the AI didn't spit it back out to us the way we needed it. We may need to provide better directions on the output.

When students reword their questions, it isn't just about getting a better answer... It's about improved communication and engaging in an iterative process. Each time adjustments are made, a student improves how they ask questions, how they think critically about answers, and refining their enquiry skills.

Adjusting the Output

As important as every part of the CLEAR check is, the ability to adjust and limit the output format is essential. As conversations unfold with AI thought partners, learning to specify and control the output format can be what turns mediocre into marvelous.

Consider this example:

- Basic: "Explain the water cycle"

- Defined output: "Explain the water cycle using the following format:
  - First, define each stage in 1-2 sentences
  - Then, create a numbered sequence of steps
  - Finally, provide one real-world example for each stage"

When students specify the format in which they want the AI to respond, they not only narrow down the possible outputs but also ensure they receive information in the most useful and clear way. Whether it's asking for a paragraph, a bulleted list, a table, or even a limerick, learning to direct the format of the response is key to maximizing the effectiveness of AI tools.

How to Teach It:

1. **Introduce Output Types**: Explain that students can ask the AI to present information in specific formats depending on what they need. Provide examples of how different formats serve different purposes:

   - **Tables**: Use tables to organize complex data or compare information across categories. This format is especially useful when students need to break down information for side-by-side analysis.

Example: "Create a table comparing renewable energy sources like solar, wind, and hydroelectric based on efficiency, cost, and environmental impact."

- **Bullet Points**: Perfect for summarizing key ideas, providing a checklist, or offering a quick overview. Bullet points help break down information into digestible chunks. Example: "List five major reasons for the start of the Civil War in bullet points."

- **Paragraphs**: Best for providing detailed explanations or narratives. A well-structured paragraph allows for more depth in addressing a topic or question. Example: "Write a paragraph explaining the main causes of the Industrial Revolution and how they contributed to societal changes."

- **Creative Formats**: Encourage creativity by asking students to generate content in formats like poems, limericks, or even short stories. This can help with retention and engagement. Example: "Write a limerick explaining the process of evaporation in the water cycle."

- **Business Formats**: Challenge students to think in professional contexts by creating outputs like newsletters, emails, or presentations. This format reinforces clarity and professionalism in communication. Example: "Draft an email summarizing the main findings of a report on climate change solutions, aimed at a corporate audience."

- **Social Media Posts**: Push students to distill complex information into engaging, concise posts. This format can develop their ability to summarize and appeal to specific audiences, combining relevance with creativity. Example: "Create a social media post summarizing the causes of World War I, with a hook and hashtags to engage a general audience."

2. **Step-by-Step Practice**: Start by having students craft a prompt that does not specify an output format, then compare it to a prompt where the output is clearly defined. Ask them to analyze the differences in clarity and usefulness between the two responses. Example:

- Prompt without output: *"Explain renewable energy"*

- Prompt with output: *"Create a table comparing the advantages and disadvantages of solar, wind, and hydroelectric power."*

As a twist on this, have students ask the same question and change the output to see what variety they get. Suddenly, we're asking AI to explain renewable energy with a poem, as a table, or as a social media post to see how that changes the perspective and details of the results.

3. **Prompt Refinement Exercise**: Have students revise their prompts by adding specific output formats. This could be as simple as asking for a "paragraph explanation," or more complex, such as requesting a "table that outlines pros and cons."

4. **Reflect on Effectiveness**: After students receive their AI responses, ask them to reflect on how effectively the AI adhered to the requested format. If the output wasn't as expected, guide them through tweaking the prompt to be even clearer about the desired structure.

This part of engineering a better prompt is not only about getting more control over what AI spits out by defining the output format, it's also about making their interactions with AI more efficient. And this is something you should also practice and model as you push your own proficiency with LLMs. They are powerful tools in the workplace. Skills like this mirror real-world communication, where clarity in requests and output format are essential for success. Plus, it reinforces critical thinking as students decide which format best serves their enquiry.

By putting all of these pieces together, we can help students become better prompt engineers, more effective users of AI tools, and digitally literate problem-solvers.

CLEAR Check Activities

Here are a few more activities to help your students target the CLEAR check and work towards refining their questioning skills.

✓ **Classroom Strategy: Prompt Puzzlers**
How it Works: The teacher posts a complex AI-generated answer. Students must analyze the answer, identify what the original question might have been, and then recreate it by crafting a prompt that would have led to that response. After testing their prompt in an AI model, students must refine it twice for better clarity or specificity, using the CLEAR check.

The Goal: To encourage students to think critically about the relationship between questions and answers, and to practice refining their enquiry to get more accurate or specific responses from AI.

Notes: This activity helps build critical thinking and AI literacy. It's great for teaching students the importance of clear and focused questioning. Teams can compete for the best, most accurate recreated question.

✓ **Classroom Strategy: AI Detective**
How it Works: In this activity, students receive an ambiguous or vague answer from AI. Their task is to break down the answer and reconstruct the question in a way that would lead to more targeted and relevant information. After testing the original question, they refine it in two rounds, with teacher guidance as needed.

The Goal: To help students understand how AI interprets vague questions and to practice refining questions to get better, more useful results.

Notes: Highlight the Explain and Reword steps of the CLEAR check process as essential for identifying gaps and improving clarity. Encourage students to collaborate and share their refined questions to compare results.

✓ Classroom Strategy: AI Interviewer

How it Works: In this role-playing activity, students take turns being the "AI interviewer" and the "AI respondent." One student asks a question to the AI model, and the "respondent" gives a short AI-generated response. The student acting as the interviewer must then adjust their question twice to get a better, more informative answer, following the CLEAR check.

The Goal: To help students practice questioning strategies in a conversational way, learning to refine prompts based on the answers they receive.

Notes: This activity enhances students' ability to think on their feet and adjust quickly. It also teaches the value of effective communication and allows for interactive, real-world practice.

These activities will help your students become better at asking questions and getting the best answers from AI. They'll learn to think critically, work together, and keep trying until they get the information they need.

It's a fun way to help them become more confident and skilled learners.

> *For a fun activity focused on refining skills, try the Reverse Image Prompt Engineering activity in the 'AI Powered Activities for Kids', located in Chapter 7 of this book or snag a copy of one that is all about pets at www.peaceframework.com/imageactivity*

RECAP: CLEAR Check

The CLEAR check is a transformative approach to refining AI interactions, turning initial frustrations into productive conversations. It emphasizes the importance of iterative questioning, encouraging users to engage deeply with AI responses rather than settling for the first answer. The steps break down to:

- **(C) Complete:** Did the AI output completely answer all parts of the initial question?
- **(L) Logical:** Is the AI output logical and directly connected to the question's intent?
- **(E) Explain:** Do you need the AI to explain or defend its output to identify gaps or misunderstandings?

231

- **(A) Adjust:** Could the AI's output be presented in a more useful format, like a table, image, or chart?
- **(R) Reword:** How can the question be rephrased or clarified to guide the AI toward a better response?

The more students practice the CLEAR check, the better they'll become at getting the information they need in a way that works best for them. This method is a constant reminder that great results come from iteration and active conversations with the AI. Effective questioning isn't about asking once and being done with it. It's about learning, adapting, and improving with each step of the process.

SAFE and CLEAR Check in the Analyze Phase

Let's review how these two ideas work together as part of the Analyze phase. By the time students are in this phase, their curiosity has been sparked, their questions refined using the BID and ReBID routines through targeted elements, and their enquiry SHARPened into a question ready for investigation. We take that question and input it into our AI chatbot or LLM to gather information about our wonder - but what happens after the AI provides a response?

This is where the **SAFE** and **CLEAR** checks become essential. While each serves a distinct purpose, they work best in tandem, guiding students to evaluate the AI outputs critically and refine their questions for iterative improvement.

Here's the simplest way to think about it:

- **SAFE** is about evaluating the response. It ensures students can critically assess the AI's output by examining the **Source**, **Accuracy**, **Fairness**, and **Evidence** of the information provided.
- **CLEAR** is about determining whether the AI's response fully meets the intent of the question. If it doesn't, CLEAR provides a structured approach for analyzing the response and refining the interaction. Students evaluate whether the output is **Complete**, **Logical**, and well-**Explained**, then decide how to **Adjust** the format or **Reword** their question to guide the AI toward better results.

Together, these two checks form a powerful feedback loop. SAFE empowers students to evaluate outputs critically, ensuring trustworthiness and quality, while CLEAR helps them improve their enquiry process, turning AI interactions into opportunities for deeper learning and clearer communication.

SAFE and CLEAR in Tandem

Let's connect these two checks together through a real-world example.

Suppose a student asks, "What are the best ways to reduce energy costs in schools?" AI provides an answer like this:

"Schools can reduce energy costs by using LED lighting, installing solar panels, and turning off electronics when not in use."

Step 1: SAFE Evaluation
Students use SAFE to assess the response:

- **Source:** "Where did the AI pull this information from? Are these credible recommendations, or is something missing?"
- **Accuracy:** "Does this align with known practices? What do I know about the cost-effectiveness of solar panels for schools?"
- **Fairness:** "Does this response consider schools in different economic contexts or climates?"
- **Evidence:** "Did the AI cite studies or examples to back up these suggestions?"

234

Through **SAFE**, students might discover that the AI's suggestions lack depth, particularly in addressing cost barriers for underfunded schools.

Step 2: Using CLEAR for Refinement
Next, students turn to CLEAR to refine their question and better engage with the AI:

- **Complete:** "Did this response fully address how schools with limited budgets can reduce energy costs?"
- **Logical:** "Is the response logical and practical for schools in different contexts? Does it align with my intent to focus on low-cost solutions?"
- **Explain:** "AI, why did you prioritize solar panels in your response? Are there other options more suited for underfunded schools?"
- **Adjust:** "Would it be helpful to adjust the output, like with a cost chart to compare solutions?"
- **Reword:** "What's the best way to rephrase my question to clarify that I'm looking for budget-friendly solutions?"

With this adjustment, the revised question becomes: *"What are cost-effective strategies for reducing energy costs in schools with limited budgets?"*

Through **SAFE**, students evaluate the initial response for its credibility, accuracy, and depth. Then, through **CLEAR**, they refine and re-engage with AI, adjusting their query to ensure a more relevant and complete answer. This systematic approach gives students a way to consistently analyze the results of their AI-powered enquiry. The questions, 'Is it SAFE?' and 'Is it CLEAR?' help maintain focus, promoting critical thinking and problem solving.

But there's one more thing to consider: It's easy to get lost in deep AI conversations. We need one last skill in the Analyze phase to stay successful - **time management**. This skill ensures that the analysis stays focused, purposeful, and productive. Let's take a deeper look.

Time Management Strategies

It's time to tackle the sneaky culprit behind most incomplete projects - poor time management. Research rabbit holes can be fascinating (and bunnies are so cute!), but students need to learn when to dig deep and when to resurface. Sometimes done is better than perfect… especially when the bell is about to ring.

To support students in managing their research effectively, introduce structured time management strategies and tools. Here's how you can help:

- Guide students in creating research timelines or project plans that break their work into manageable chunks.
- Provide easy-to-use templates or checklists for tracking progress, milestones, and deadlines.
- Teach techniques for prioritizing tasks and allocating time efficiently.
- Introduce productivity tools or apps that encourage accountability and focus.

AI tools are great companions here, serving as personal assistants to help students organize and optimize their work. Share and review prompts like the ones below to set students up for success.

Time Management Prompts

These prompts are differentiated for younger and older students:

Project Planning and Prioritization

- *Younger Students*:
  "Make a checklist of steps to follow so I can (outcome or project). It is due on (date)."
- *Older Students*:
  "Create a detailed project plan for my (project name) assignment, breaking it down into smaller tasks with deadlines to keep me on track. It's due on (date)."

Time Tracking and Analysis

- *Younger Students*:
  "What time do I need to finish my homework so I still have time to play?"
- *Older Students*:
  "Analyze my daily schedule and suggest ways to improve my time management so I can finish my assignments and still have time for other activities."

Stress Management and Productivity

- *Younger Students*:
  "What are some fun ways to stay calm when I feel worried about finishing my work?"
- *Older Students*:
  "Suggest mindfulness or productivity techniques

238

to help me stay focused and calm when I'm stressed about deadlines or too much work."

Goal Setting and Achievement

- *Younger Students*:
  "Help me make a simple plan to help me (goal, e.g., finishing my book, learning my spelling words)."
- *Older Students*:
  "Create a personalized study plan that helps me achieve my academic goals by breaking them into weekly tasks."

Time management isn't just for the classroom; it's a lifelong skill. Whether students are managing research projects, studying for finals, or juggling multiple responsibilities in the workplace, these strategies prepare them to succeed in an AI-powered world.

With the right tools, templates, and prompts, students can avoid feeling overwhelmed and confidently tackle complex, open-ended enquiry projects. And the bonus? They'll build the self-discipline and organizational skills necessary in any deadline-driven environment.

Let's teach them how to manage their time, so they can make the most of it... rabbit holes and all.

In a Nutshell: Analyze

Remember where we started? In the *Provoke* phase, we sparked curiosity. During *Enquire*, we fine-tuned our questions. Now, in *Analyze*, students are actively interrogating for information. They're critical of the information they receive, comparing sources, and applying critical thinking to evaluate what they find. This is where we use the SAFE check to arm them with the skills they need to navigate the learning process and differentiate quality information from noise.

Then, as they work with LLMs to answer those burning questions, they converse with AI using the CLEAR check to determine how to move forward. Imagine students working on a research project about ocean ecosystems. They're not just Googling for quick answers; in Analyze, they're evaluating information, organizing timelines to stay on track, and using a conversational approach to get answers in a format that is helpful to them and their learning.

Here's how this phase might look in practice:
240

"We have our focused question to prompt AI. Let's start the enquiry process, and remember to use the SAFE check to evaluate your sources. Take a minute to ask yourself: Is this data credible? Relevant? Accurate? What's missing, and why might that matter?"

"I want to pause for a moment here. You should be iterating your prompts to get better and more helpful information. As you ReBID, remember the CLEAR check. Is the AI response complete? Logical? Well-explained? Are you adjusting the output or rewording your query for better results? Let's check in with a partner and help each other update your prompts."

"Back together now. I hope that 'think' time with your partner helped you come up with better ways to get what you want from your AI tool. As you keep working on your enquiry question, continue to check if the results are SAFE and CLEAR."

"You should have some really quality information that helps you think about your enquiry question. We're looking for insights here, not just answers. It's time to meet with a partner, share some of the juiciest bits of information you found, and explain how you think it supports your enquiry question."

"Now that we've collaborated, let's talk about a deadline date so you can set a timeline and get ready to create."

Here's the pieces we are putting together:

- **Be Critical About Sources:** Reviewing source credibility, accuracy, fairness, and evidence helps students know if they are using **SAFE** information, whether it's AI-generated or traditional.
- **Evaluate Results:** The **CLEAR** check guides students to refine questions and adjust their approach for better outcomes.
- **Manage the Research Process:** Time management strategies ensure students can balance deep analysis with efficient workflow.

By the end of Analyze, our students aren't just skimming for facts or clicking on Google's top result... they're transforming into thoughtful data analysts, capable of distinguishing reliable information from the rest and adjusting their query by better understanding AI outputs and responses. Plus, they'll understand that analysis isn't just about getting answers; it's about finding *the right answers for the right reasons*.

242

In the Analyze phase, we're not just guiding students to use AI; we're teaching them how to think *with* it, setting them up with skills they'll rely on long after they've left our classrooms.

Next up: The Create phase, where we put all this new information to good use!

***

Create

Now that we've provoked curiosity (P), enquired through questioning (E), and analyzed information (A), it's time to let students flex their creative muscles (C)! Remember that time you assigned a "simple" creative project and somehow ended up with one student building an entire solar system from papier-mâché while another handed in a hastily drawn stick figure? (We've all been there!) The Create phase is about finding that sweet spot between unleashing creativity and maintaining manageable expectations. It's where students transform their analyzed information into something meaningful - without driving us to a caffeine overdose trying to support 30 different project types.

Opportunities for Creation

When it comes to enquiry-based learning, the types of artifacts students can create are limited only by their imagination. While traditional reports have their place, they can be easily generated with AI tools, which is why we should encourage students to think outside the box. The goal is to make learning tangible and engaging. Students should apply their knowledge creatively, rather than just memorizing facts. Here are some realistic ideas for K12 teachers that tap into creativity and allow students to be innovative.

Creating in the Early Years (Lower Elementary)

1. **Visual Explanations**
   - **Non-Tech:** Create posters or flip books using drawings and simple labels to explain concepts.
   - **Tech:** Use simple drawing apps or slide presentations with voice recordings to explain ideas.

2. **Show and Tell Presentations**
   - **Non-Tech:** Create display boards with pictures and artifacts to share their learning.
   - **Tech:** Record short video presentations using tablet devices or create simple digital slideshows.

3. **Learning Journals**
   - **Non-Tech:** Maintain illustrated notebooks with drawings, cutouts, and simple writing.
   - **Tech:** Build digital portfolios using photos and voice-to-text features.

4. **Storytelling**

- **Non-Tech:** Students create storybooks using paper and crayons, illustrating their narratives.
- **Tech:** Use book and comic creation apps to digitally craft and publish stories.

5. **Art and Music**
- **Non-Tech:** Students paint pictures, create sculptures, or compose songs to express their understanding.
- **Tech:** Use digital art tools like Paint or Canva to create visual representations of their learning.

6. **Digital Creations**
- **Non-Tech:** Students can use simple drawing tools or building blocks to create physical models.
- **Tech:** Use coding platforms to create interactive stories, games, or animations.

Creating in Upper Elementary

1. **Educational Guides**
- **Non-Tech:** Create how-to booklets or instructional posters with step-by-step guidance.

- **Tech:** Design digital tutorials using screencasts or slide presentations.

2. **Information Displays**
   - **Non-Tech:** Design tri-fold boards or wall displays showcasing research findings.
   - **Tech:** Create interactive digital boards using tools like Padlet or Figma.

3. **Creative Summaries**
   - **Non-Tech:** Develop comic strips or illustrated timelines.
   - **Tech:** Create digital infographics or animated explanations.

Creating in the Middle Years (Upper Elementary and Middle School)

1. **Interactive Timeline**
   - **Non-Tech:** Create a physical timeline using a long piece of paper or a bulletin board.
   - **Tech:** Use digital timeline tools to create interactive timelines.

2. **Social Media Campaign**
   - **Non-Tech:** Create posters, flyers, or brochures to promote a cause or share an idea.

- **Tech:** Create digital materials appropriate to share information and engage with others in a consumable way (TikTok not required).

3. **Design Challenges**

- **Non-Tech:** Build physical models or prototypes using materials like cardboard, paper, or recycled items.
- **Tech:** Use CAD software or 3D printing technology to design and create innovative solutions.

4. **Experimentation and Enquiry**

- **Non-Tech:** Conduct simple experiments using household materials.
- **Tech:** Use data logging devices or online simulation tools to collect and analyze data.

5. **Podcasting**

- **Non-Tech:** Be interviewed by a fellow student about a topic, idea, or question in front of the class.
- **Tech:** Use recording software to record and edit an episode.

Creating in the Upper Grades (Middle and High School)

1. **Multimedia Presentations**
   - **Non-Tech:** Design presentation boards with rotating elements or flip-up features.
   - **Tech:** Create narrated slideshows or interactive digital presentations.
2. **Documentary Style Projects**
   - **Non-Tech:** Create newspaper articles or magazine spreads.
   - **Tech:** Produce video documentaries or podcast episodes.
3. **Teaching Tools**
   - **Non-Tech:** Design board games or card games that teach concepts.
   - **Tech:** Create digital quizzes or interactive learning activities.
4. **Science Fair Projects**
   - **Non-Tech:** Conduct experiments and present findings using tri-fold boards.
   - **Tech:** Use data analysis tools and presentation software to create professional-looking exhibits.
5. **Photo Essays**

- ○ **Non-Tech:** Take photographs and arrange them in a physical photo album to explain an idea or share information.
- ○ **Tech:** Use photo editing software to enhance images and create digital photo essays.

6. **Role-Playing Scenarios**
   - ○ **Non-Tech:** Act out scenes in front of the class or another live audience.
   - ○ **Tech:** Use video editing software to create short films or documentaries.

7. **Graphic Novels**
   - ○ **Non-Tech:** Create comic strips using paper and pencil.
   - ○ **Tech:** Use digital comic book creation tools to design and publish graphic novels.

8. **Public Service Announcements**
   - ○ **Non-Tech:** Create posters, flyers, or scripts appropriate for radio, social media, or TV PSAs.
   - ○ **Tech:** Use design or video editing software to produce high quality PSAs.

9. **Art Installations**
   - ○ **Non-Tech:** Create physical installations, such as paintings and sculptures, using traditional art materials.

o **Tech:** Use digital art tools or projection mapping to create immersive installations.

10. Interactive Games

o **Non-Tech:** Design board games or card games.

o **Tech:** Use game development software to create digital games.

11. Debate Preparation Kits

o **Non-Tech:** Research and organize arguments using traditional note-taking methods.

o **Tech:** Use online text documents to prepare arguments and counterarguments for or against an issue.

12. Music Composition

o **Non-Tech:** Compose music using traditional instruments or vocal techniques.

o **Tech:** Use digital audio workstations to create and produce music.

13. Historical Fiction Stories

o **Non-Tech:** Write short stories or novels using pen and paper.

o **Tech:** Use word processing software to write and edit stories.

14. Data Visualization Projects

- ○ **Non-Tech:** Create charts and graphs by hand or using basic spreadsheet software.
- ○ **Tech:** Use data visualization tools like Tableau or Power BI to create interactive visualizations.

15. **Thematic Scrapbooks**
- ○ **Non-Tech:** Collect and organize physical items like newspaper clippings, photographs, and postcards.
- ○ **Tech:** Create digital scrapbooks using online tools or presentation software.

Creating in High School

1. **Analysis Products**
   - **Non-Tech:** Create analytical essays with visual evidence or debate preparation cards.
   - **Tech:** Develop digital analysis dashboards or video essays.
2. **Research Showcases**
   - **Non-Tech:** Design research posters or investigation journals.
   - **Tech:** Create digital research websites or interactive data presentations.
3. **Problem-Solution Projects**

- **Non-Tech:** Develop proposal packets, solution blueprints, or contingency maps.
- **Tech:** Create digital prototypes or simulation models.

4. **Debates and Discussions**
   - **Non-Tech:** Students can participate in face-to-face debates or write persuasive essays.
   - **Tech:** Use online forums or social media to engage in digital debates and discussions.

5. **Persuasive Writing and Advocacy**
   - **Non-Tech:** Students can write letters, create posters, or give speeches.
   - **Tech:** Use digital tools to create multimedia presentations, infographics, or videos.

6. **Original Research and Analysis**
   - **Non-Tech:** Students can conduct library research and analyze primary sources.
   - **Tech:** Use online databases and statistical software to analyze data and draw conclusions.

7. **DIY Inventions**
   - **Non-Tech:** Build physical prototypes using simple tools and materials.
   - **Tech:** Use 3D printing or laser cutting to create more complex inventions.

8. **Film Documentaries**
   - **Non-Tech:** Film and edit videos using basic video editing software.
   - **Tech:** Use professional-grade video editing software to create high-quality documentaries.

The possibilities are endless here. So which one do you choose? Whatever your students are into! You can even use a choice board here to let students drive their own creation direction within your preset parameters. Not sure what your students are into? Try a survey at the start of a unit. This can help you pick project options perfect for their interests.

Making Meaning (Without Losing Our Minds)

Remember those solar system models we made as kids? You know, the ones with foam balls painted like planets, hanging precariously from wire coat hangers? Or those sugar cube pyramids that somehow always ended up missing a few cubes? (We all know where those went!) Here's the thing: Those projects weren't just busy work... They were our first attempts at demonstrating understanding through creation.

Let's address the elephant in the room: When teachers hear "project-based learning" or "creative

254

assessment," some of us break out in a cold sweat. We imagine elaborate schemes requiring supplies we don't have, time we can't spare, and patience we ran out of last October. But here's the beautiful truth about the Create phase of PEACE: It's not about reinventing the wheel; it's about letting students take the wheel of their own learning journey.

Why This Looks Familiar (In a Good Way)

Those tried-and-true project formats we listed? They're intentionally familiar. Making a poster about photosynthesis or recording a video about the water cycle isn't revolutionary, I know, but it works. The magic isn't in the format; it's in the journey that got us here. When students have properly progressed through Provoke, Enquire, and Analyze, the Create phase becomes less about the glitter glue and more about genuine understanding.

It's All in the Prep (That's the Secret Sauce)

Think of it this way: Those earlier PEACE phases are like prep work in the kitchen. When you've properly measured, mixed, and prepared your ingredients, the actual baking is the easy part. Similarly, when students have:

- Had their curiosity sparked (Provoke)
- Generated meaningful questions (Enquire)
- Done thorough research and analysis (Analyze)
  ...the creation part flows naturally, whether they're using construction paper or coding an app.

Breaking Free from Project Paralysis

Here's what trips us up most often: We think every project needs to be a show-stopper. Let's get real here, sometimes a simple sketch explaining cell division can demonstrate more understanding than a three-week diorama extravaganza. The key is flexibility and purpose. Ask yourself:

- Does this creation allow students to demonstrate their understanding?
- Can it be completed with reasonable resources?
- Does it align with our learning objectives? If you can answer yes to these, you're golden.

Technology: The Great Enabler (Not the Main Event)

Sure, we've included tech options in our list. But notice something? The core ideas remain the same whether students are using markers or tablets. As new tools emerge, they'll simply provide different ways to demonstrate the same understanding. The medium isn't the message - the learning is.

Prep for Diverse Learning Needs

It is on us as the teachers to make sure every student can shine in the 'Create' phase. To do that, we need to tailor our projects and make sure everyone has the tools they need.

For students who might struggle with organization or writing, consider templates, checklists, or graphic organizers - these concrete supports can make all the difference. Need more visual aids? Try adding task analysis strips, navigation icons, or clear models of both examples and non-examples.

And let's not forget about accessibility. Whenever possible, materials and resources should be accessible to everyone, including students with disabilities. As you design digital content and enquiry-based activities, think about how to incorporate assistive, instructional, and everyday technologies that improve access. This could mean providing audio versions of texts, using predictive text programs, or explicitly teaching the accessibility tools built into classroom devices.

When we do all this, we create a learning space where everyone feels included and can show off their skills. The point is this: Just as we adapt classroom activities to meet diverse needs, we should also adapt projects through target accommodations and individualized supports. Avoid excluding students with the thought, "It's too hard for them." Every student can create and contribute meaningfully with flexible options and targeted support.

The Bottom Line on Projects

The Create phase isn't about producing museum-worthy masterpieces or viral TikTok sensations. It's about giving students space to demonstrate their learning in ways that make sense to them. Sometimes that's a paper poster, sometimes it's a digital presentation, and yes, sometimes it might still be a foam ball solar system. And you know what? That's perfectly okay because when students have been engaged in meaningful enquiry throughout the PEACE process, the creation phase is *A* step, not *THE* step.

Remember: Keep it simple, keep it meaningful, and keep your sanity. The best projects are often the ones that don't require a trip to three different craft stores or a PhD in computer programming to complete.

Here's where we come full circle. Yes, we've used AI tools, engaged in deep enquiry, and potentially created some non-traditional products. But we can still assess these creations using familiar standards:

- Does it demonstrate understanding of key concepts?
- Is the information accurate and well-organized?
- Can the student explain their thinking?
- Has learning growth occurred?

We can even use LLMs to create a more robust grading rubric with key guidelines and benchmarks based on the specific creation project students undertake. Or we can use these 4 metrics to establish a standardized way to measure learning and evaluate a student's creation. Leave the idea that this has to be more complicated behind you. It's just this simple.

Classroom Vignette: Making Time for Creation

Mrs. Dell's 2nd-grade classroom hums with purposeful activity as students rotate through learning stations. With only two classroom tablets, she's found a way to blend traditional curriculum with creative enquiry work.

The Setup

The station rotation schedule is displayed on a colorful wall chart:

- Reading Circle
- Math Games
- Writing Workshop
- Creation Station
- Teacher Time

Strategic Implementation

Morning Stations (45 minutes):

- 5 students: Guided reading with teacher
- 4 students: Math fact practice with manipulatives
- 4 students: Writing prompt work
- 4 students: Project creation time
- 4 students: Independent reading

Station Features:

Creation Station:

- Project materials organized in clear bins
- Simple AI prompt cards for inspiration
- Voice recorder for oral storytelling
- Partner feedback protocols

Traditional Learning Stations:

- Self-checking materials
- Clear success criteria
- Peer support roles
- Quick assessment tools

Sample Student Schedule

Carson's Monday Morning:

1. 8:30-9:15: Math Games Station
   - Completes required practice
   - Records progress in math journal
2. 9:15-10:00: Creation Station
   - Works on habitat diorama
   - Uses AI prompt card for animal facts
3. 10:00-10:45: Teacher Time
   - Reading group instruction
   - Quick skill check

Management Tools:

- Visual timer for transitions
- "Ask 3 Before Me" rule
- Success criteria checklists
- Quiet signal system
- Partner progress checks

Carson's Share Circle Comment:

"I love Creation Station time! It's my favorite part of the day because I get to work on my special project. And I like the AI cards because they help me think of new ideas when I get stuck. Yesterday, I learned that polar bears have black skin under their white fur! I put that fact in my habitat project, and Emma, my partner, thought it was super cool. I like that I don't have to rush through my project, and Mrs. Dell helps us too. Sometimes I finish my math games really fast so I can have extra time at Creation Station!"

Teacher Reflection

"The station rotation model is amazing, especially in managing my fast finishers. Students get focused instruction in core subjects while still having dedicated time for creative projects. The structure actually gives us more flexibility - when students finish station work early, they can seamlessly move over to their project work. Even with our limited technology, we're able to integrate AI through simple prompt cards and recorded responses. The kids are more engaged, plus I'm able to provide better support because everyone knows exactly what they should be doing at each station."

Assessments with Enquiry-Based Learning

Remember when "assessment" meant collecting 30 identical worksheets and grading them with your trusty red pen? (Those were simpler times, weren't they?) But here's the thing: Just because our teaching has evolved doesn't mean we have to throw out everything we know about assessment. In fact, the PEACE Framework gives us more opportunities (not fewer) to measure student growth and understanding.

Let's start with that pesky elephant in the room again… state assessments aren't going anywhere. But here's the good news - when we set clear expectations from the beginning, we're actually preparing students for both creative enquiry AND standardized success. Let's work through how to make that happen.

The Assessment Overlaps You Didn't Know You Needed

Look, I get it. The moment someone suggests changing how we assess student learning, most of us instinctively clutch our grade books a little tighter. After twenty years in education, I still start chewing on my pen cap when some hooha starts talking about "change" and "alternative assessments". *But what about the standards? The district benchmarks? The parents expecting traditional grades?*

Take a deep breath. We're not throwing out the grade book - we're just reorganizing it. And here's the plot twist: This might actually make your life *easier*.

Map Your Measurements

Let's say we were teaching a unit about weather patterns. Traditionally, we'd give a spelling test, a couple of quizzes or a unit test, and some worksheets on understanding a precipitation chart or temperature graph. These three red-pennable moments would give us a grade that indicates if a student 'got it'.

Let's map this out in a different way that might just make you fall in love with enquiry-based assessment.

(Okay, maybe "fall in love" is a stretch, but at least not break out in hives.)

Traditional Assessment Points

Here are the assessment points in a traditional approach; vocabulary, comprehension, and analysis. Here is the underlying knowledge or skill we hope these assessment points capture.

1. **Vocabulary Mastery**
   - Weather-related terms
   - Scientific processes
   - Tool/instrument terminology
2. **Concept Comprehension**
   - Understanding weather systems
   - Cause and effect relationships
   - Pattern recognition
3. **Data Interpretation**
   - Reading weather maps
   - Analyzing temperature data
   - Interpreting precipitation patterns

PEACE Framework Checkpoints

Now, let's think of how we would likely assess students through enquiry-based learning with a breakdown of the underlying skills or knowledge.

1. **Enquire Phase Assessment**
   - ○ Quality of student questions
   - ○ Depth of enquiry
   - ○ Connection to prior knowledge

2. **Analyze Phase Assessment**
   - ○ Research accuracy
   - ○ Source evaluation
   - ○ Data interpretation skills

3. **Create Phase Assessment**
   - ○ Concept application
   - ○ Innovation in presentation
   - ○ Depth of understanding

The Magic Intersection

Here's where it gets good: These aren't separate assessment tracks - they're the *same things viewed through different lenses*. Let me show you how:

1. Vocabulary Mastery → Enquire Phase

Instead of giving a traditional vocabulary quiz, assess vocabulary naturally through:

- Student-generated questions (Are they using terms correctly?)

- Research notebooks (Are they incorporating instructional vocabulary?)
- Discussion contributions (Can they explain concepts using proper terminology?)

2. Concept Comprehension → Analyze Phase

Rather than a multiple-choice test, evaluate understanding through:

- Research summaries (Do they grasp core concepts?)
- Data collection methods (Are they looking for the right things and getting the desired results?)
- Peer explanations (Can they teach others?)

3. Data Interpretation → Create Phase

Instead of worksheet-based analysis, assess through:

- Student-created visualizations
- Real-world applications
- Problem-solving scenarios

Ultimately, this integrated approach benefits students by providing them with multiple opportunities to demonstrate their knowledge and skills. This not only enhances student learning but also streamlines the assessment process. By aligning traditional assessments with enquiry-based learning, we can maximize our time and effort and create a holistic picture of student learning.

Making It Work (Without Losing Your Mind)

Here's your sanity-saving strategy:

1. **Use AI as Your Assessment Assistant**
   - Let AI help track vocabulary usage in student work
   - Generate quick comprehension checks during discussions
   - Help analyze student questions for depth and relevance
2. **Create Multi-Purpose Assessment Opportunities**
   - Design projects that naturally incorporate vocabulary, concepts, knowledge, and skills
   - Use discussion protocols that reveal understanding while building skills
   - Implement peer review systems that reinforce learning while providing assessment data
3. **Build Assessment into Daily Practice**

- Use exit tickets that align with both traditional and PEACE Framework goals
- Implement digital portfolios that track progress across all metrics
- Create rubrics that satisfy both assessment approaches

Classroom Vignette: Creating with Purpose

In Mrs. Vine's 10th-grade World History class, students are transitioning from research to creation for their "Revolution Through Time" projects. The energy is high, but so are her concerns about maintaining academic integrity in an AI-rich environment.

The Setup

Students have investigated historical revolutions of their choice through the PEACE Framework and are ready to demonstrate their understanding through creative artifacts.

Assessment Strategy in Action

Clear Expectations:

- Digital rubric shared with specific criteria:
    - Historical accuracy (40%)
    - Creative interpretation (30%)
    - Process documentation (20%)
    - Peer collaboration (10%)
- Sample projects from previous years (AI and non-AI examples)

Creation Options:

- Interactive timeline with AI-generated visuals
- Historical figure dialogue podcast
- Revolution comparison infographic
- Documentary-style video essay
- Choose-your-own-adventure historical simulation

AI Integration Checkpoints

Monday: Project proposal with AI brainstorming documentation

Wednesday: Progress check-in with AI usage log

Friday: Peer review using AI-assisted feedback protocol

Sample Student Journey: Charlie's Revolutionary Radio Show

1. Initial Planning
   a. Uses AI to generate interview questions for historical figures
   b. Documents which questions are AI-generated vs. original
   c. Submits planning reflection via quick-write
2. Development Phase
   a. Creates script incorporating primary sources
   b. Uses AI for language authenticity check
   c. Completes mid-point self-assessment
3. Final Production
   a. Records dialogue with period-appropriate details
   b. Includes process portfolio showing evolution
   c. Presents to class with Q&A session

Alex's Project Reflection:

"Creating the Revolutionary Radio Show pushed me out of my comfort zone, but in a good way. Brainstorming with AI about interview questions for historical figures was, like, the best. It came up with questions I would have never thought to ask. It was hard deciding which questions would actually tell me the most interesting stuff about the revolution. I actually really liked documenting my AI use because it made me think more carefully about each choice I made (I didn't want Mrs. Vine thinking I used it too much). Some of my best ideas came from disagreeing with the AI's suggestions and then figuring out why I thought differently. My favorite part was when my classmates said they finally understood why the French Revolution was more than just angry peasants after listening to my radio show."

Teacher Reflection

"The key was making the process as important as the product. By requiring students to document their AI usage and thinking process, I could see their learning journey clearly. The rubric's emphasis on original thinking alongside technical execution helped students use AI as a tool rather than a shortcut. Best of all, their excitement about the projects made the assessment feel more like a celebration than an evaluation."

Universal PEACE Framework Rubric

This base rubric can be adapted for any subject area

Enquire Phase (12 points)

Criteria	Beginning (1 pt.)	Developing (2 pts.)	Proficient (3 pts.)
Question Quality	Questions are basic or surface-level	Questions show some depth but may lack focus	Questions demonstrate critical thinking and relevance
Prior Knowledge Connection	Limited connection to existing knowledge	Some connections made to prior learning	Clear connections to prior knowledge and experiences
Research Planning	Basic or incomplete research plan	Adequate research plan with some structure	Well-structured research plan with clear goals
Vocabulary Usage	Limited use of content vocabulary	Some appropriate use of content vocabulary	Consistent and accurate use of content vocabulary

Analyze Phase (12 points)

Criteria	Beginning (1 pt.)	Developing (2 pts.)	Proficient (3 pts.)
Source Evaluation	Uses limited or unreliable sources	Uses some credible sources	Uses multiple credible sources effectively
Data Analysis	Basic data interpretation with errors	Adequate data interpretation with minor errors	Accurate data interpretation and analysis
Pattern Recognition	Identifies simple patterns	Recognizes patterns and some relationships	Identifies complex patterns and relationships
Critical Thinking	Basic analysis with limited depth	Some critical analysis present	Clear critical thinking and analytical skills

Create Phase (12 points)

Criteria	Beginning (1 pt.)	Developing (2 pts.)	Proficient (3 pts.)
Content Knowledge	Basic understanding shown	Adequate understanding demonstrated	Clear and complete understanding
Innovation	Standard presentation approach	Some creative elements present	Creative and engaging presentation
Technical Skill	Basic use of tools/media	Competent use of tools/media	Skilled use of tools/media
Delivery	Basic delivery of ideas	Clear delivery of main points	Effective delivery of complex ideas

Subject-Specific Quick Guides

Math Assessment Integration

Traditional Points → PEACE Framework

- Calculation accuracy → Analysis of problem-solving process

- Formula memorization → Pattern recognition in real-world applications

- Problem completion → Creation of original mathematical models

Language Arts Assessment Integration

Traditional Points → PEACE Framework

- Vocabulary tests → Usage in authentic writing
- Reading comprehension → Analysis of themes and connections
- Essay structure → Creative expression with proper conventions

Science Assessment Integration

Traditional Points → PEACE Framework

- Term definitions → Scientific language in investigations
- Concept tests → Experimental design and analysis
- Lab reports → Original research presentations

Social Studies Assessment Integration

Traditional Points → PEACE Framework

- Date/fact memorization → Historical pattern analysis
- Document analysis → Multi-source investigation
- Event summaries → Original historical interpretations

Classroom Vignette: Cross-Disciplinary Weather Unit

Mrs. Perez's 7th-grade classroom buzzes with activity as students work on their climate change investigations. What started as a science unit has evolved into a cross-disciplinary project that hits multiple traditional assessment points while following the PEACE Framework.

The Setup

Students are investigating the question: "How is climate change affecting our local community, and what can we do about it?"

Assessment in Action

Morning Session: Math & Science Integration

- Students analyze 30 years of local temperature data
- *Traditional Assessment Point:* Data interpretation, graphing skills

277

- *PEACE Framework Checkpoint:* Analyze phase - Pattern recognition, data visualization
- *Teacher Tool:* Digital portfolio entries where students explain their methodology

Mid-Morning: Language Arts Connection

- Students craft interview questions for local farmers
- *Traditional Assessment Point:* Writing mechanics, question formation
- *PEACE Framework Checkpoint:* Enquire phase - Question quality
- *Teacher Tool:* AI-assisted rubric checking for question depth and clarity

Afternoon: Social Studies Tie-In

- Students research historical land use changes
- *Traditional Assessment Point:* Research skills, historical analysis
- *PEACE Framework Checkpoint:* Analyze phase - Source evaluation
- *Teacher Tool:* Collaborative digital workspace with embedded assessment tracking

Sample Student Work Flow

Tony's Project Journey:

1. **Enquire Phase**
   - Generated questions about local temperature changes
   - *Assessment:* Used rubric to evaluate question quality
   - *Traditional Grade:* Vocabulary usage in question formation
2. **Analyze Phase**
   - Created temperature trend graphs
   - Interviewed community members
   - *Assessment:* Data analysis skills
   - *Traditional Grade:* Math calculations and graph creation
3. **Create Phase**
   - Developed interactive presentation
   - *Assessment:* Depth of understanding shown in final product
   - *Traditional Grade:* Content mastery and presentation skills

Tony's Digital Portfolio Entry:

"When we started the climate change project, I didn't think I'd be good at it because there was so much math, and I'm not that good at math. But I made my own graphs and interviewed an AI farmer... and that made everything click. Using AI to help check my interview questions made me feel more confident when doing the interview. The coolest part was when our class data all showed the same patterns, it made me feel like a real researcher! My mom was really impressed when I explained how our community's weather has changed over time using my graphs. I think I might want to be a meteorologist when I grow up."

Teacher Reflection

"Instead of giving separate assessments for each subject area, I was able to evaluate student progress across multiple disciplines through their project work. The PEACE Framework actually made it easier to track student growth while meeting traditional assessment requirements, and I didn't have to take home a stack of stuff to grade. Using electronic grading rubrics on my tablet helped make tracking student progress easy so I could focus on teaching and not just grading."

Implementation Tips

1. Start with a small cross-disciplinary unit
2. Use AI tools to track vocabulary usage and concept understanding
3. Create digital portfolios that show growth over time
4. Involve students in developing assessment criteria
5. Share rubrics early and often
6. Use quick check-ins to monitor progress
7. Celebrate both traditional and innovative achievements

Remember: The goal is to make assessment more meaningful while reducing teacher workload. The PEACE Framework helps achieve both.

Progressive Check-ins: Your Assessment Safety Net

Let's be honest - nobody became a teacher because they loved grading papers at midnight. Assessments shouldn't feel like a burden as much as a natural part of the learning journey. That's why we need progressive check-ins! Think of these like those mile markers on a hiking trail - they help you know you're heading in the right direction without slowing down your progress.

Daily Quick-Checks

- **Exit tickets**: Quick temperature checks on understanding
- **Process journals**: Where students document their thinking
- **Question logs**: Tracking how enquiry evolves

Pro Tip: Use AI tools to analyze these quick checks for patterns in understanding and vocabulary usage. It's like having a teaching assistant who is always ready to work, day or night.

Weekly Benchmarks

- **Understanding summaries**: Mini-milestone check-ins
- **Skill demonstrations**: Show-what-you-know moments
- **Standard alignment checks**: Keeping our educational GPS on track

Note: Checking in with students as they work through this process is the best way to keep them on track and address any misconceptions. Check in so students can quickly summarize key concepts in their own words for you. If you can't get to every student, use tech tools to help! "Show-What-You-Know Moments" can be incorporated into daily routines, where students briefly share their insights or answer a prompt related to the current topic. The point here is to keep your fingers on the pulse of student learning.

Students Portfolios

Remember when we used to just collect final papers and call it a day? Those were simpler times. But were they better? Not really. We need to document progress, but without needing a rolling cart to get all those papers home to grade on the sofa. As you work through the PEACE process, we have grading rubrics and a final project to assess, but you can also work on project portfolios to show evidence of learning. These can include:

- **Student-Selected Work:** Pick pieces that represent their best work, from initial drafts to final projects.

- **Student Reflections:** Include insights into the student's thinking process, their challenges, and their successes.

- **Annotated Work Samples:** Highlight specific skills or strategies used in each piece of work.

- **Self-Assessment:** Use student rubrics or checklists to evaluate their own work.

- **Peer Feedback:** Incorporate feedback from classmates to improve their work.

- **Teacher Feedback:** Include notes and feedback you've provided and how it was incorporated.

Think of it as creating a story of learning rather than just a grade.

Test Prep That Doesn't Feel Like Test Prep

No teacher wants to 'teach to the test'. The fact is it's *not* the purpose of learning, it's a function of assessment. We have other ways to assess, but you're way ahead of your time and still need to answer to Mr. Scantron. How can we cover our bases?

- **Ask to the Test:** When you formulate your focus and align with state standards, choose topics that you know will be asked on the test. It's likely a critical area for student knowledge and skills.

284

- **Let Students Write the Test:** After the create process is finished, or as one of the project options, let students write a test that they think assesses the lesson objective and includes the answer key.

- **Bank Test Questions:** As students are asking questions as part of the Enquire phase and searching for information as part of the Analyze phase, have them bank questions they think would appear on a standardized test. Then you can set aside time to review them and see if students have answers (two teams pitted against each other trying to answer the most is a great review game).

- **Research the Test:** Let students use AI as a study partner to review test questions and topic information as part of improving test results after the Create phase.

- **Teach Test-Taking Skills:** Since students aren't taking as many classroom tests, we have to teach them to dissect complex questions by identifying keywords and breaking down the parts so they are test-ready. Show students how to spot clues, understand instructional terms, or identify keywords in reading passages that hint at main ideas or themes as part of prompt engineering.

Remember, the goal isn't to force test prep into the curriculum; it's to integrate it seamlessly into the learning process. By focusing on enquiry, creativity, and critical thinking, we're helping students to succeed on the test and in life.

Flexible Feedback Loops

No one likes being judged, but we love when assessments are more about conversation focused on progress over perfection.

- **Ongoing guidance**: Real-time feedback that shapes learning, whether it's a quick comment on a rough draft or a detailed critique of a final project.
- **Specific strategies**: Give clear and actionable paths for improvement.
- **Growth opportunities**: Include multiple chances to demonstrate understanding.
- **Reflection:** Use self-assessment, peer feedback, or teacher-student conferences to reflect and identify areas for growth.

It's no longer about right or wrong, good or bad, pass or fail. It's about creating, reflecting, receiving feedback, making change, and creating again. We need to solidify this process idea with flexible feedback loops like this, because learning isn't linear.

The Bottom Line

You're not teaching twice as much… You're teaching *once*, but measuring it *well*. The PEACE Framework doesn't add new assessment requirements; it helps you gather deeper, more meaningful data about student learning while meeting your existing obligations.

Think of it like weather radar versus a thermometer. Sure, the thermometer gives you a number, but the radar shows you the whole system in motion. With the PEACE Framework, you're not just measuring isolated data points - you're tracking the entire learning ecosystem.

Pro Tip: Start small. Pick one unit where you'll parallel track both assessment systems. You might be surprised to find that the PEACE Framework actually gives you more concrete evidence of student learning than traditional assessments ever did.

Remember: The goal isn't to create more work - it's to make the work you're already doing more meaningful. And maybe, just maybe, make those assessment hives a thing of the past.

In a Nutshell: Create

Create is the most rewarding phase of enquiry-based learning for most students because they can see their enquiry come alive. By aligning these creations with clear assessment criteria and encouraging the use of AI tools, you're doing more than just teaching facts. You're blending learning, creativity, and technology in a way that's engaging, personalized, and fun. It's exciting to watch students dive into their projects with motivation and enthusiasm.

In this phase, remember that the focus is on the journey, not just the destination. After all, who doesn't want to see what incredible things their students can create when they're given the chance? Let the creativity flow!

Remember these essential elements:

1. **Focus on process, not just product:** Document the journey, celebrate growth, and value iteration.

2. **Ensure universal access:** Use multiple expression modes, varied tech options, and diverse project choices.
3. **Balance traditional and innovative:** Align with standards, incorporate test prep naturally, and build skills progressively.
4. **Use tech wisely:** Access digital tracking tools, try portfolio platforms, and consider assessment helpers programs.

Here's the beautiful truth: When we assess through the PEACE Framework, we're not just measuring learning - we're fostering it. Each checkpoint becomes a learning opportunity, each assessment a chance for growth.

And that midnight grading session? Well, it might not disappear entirely (sorry!), but it transforms into something more meaningful. You're not just marking papers - you're mapping learning journeys.

Now that's something worth staying up late for.

Next up: The Engage phase, where we'll explore how to turn all this amazing student work into powerful learning conversations...

***

Engage

Remember that science fair project you did in middle school? The one where you stood next to your tri-fold board while three people (including your mom) politely nodded? Yeah, we're not doing that anymore. Welcome to the Engage phase, where sharing goes to the next level.

Going Beyond the Project

Let's be real - if the only audience for student work is their teacher, we're missing a massive opportunity.

Here's how to transform "*show and tell*" into "*teach and inspire*":

Demonstration Strategies

Here are a few classroom-tested options that allow students to use their project as a springboard into sharing ideas and teaching others.

- **Micro-Teaching Sessions:** A 10-minute or less expert share session where students teach specific concepts to peers.

- **Micro-Learning Library:** Take those micro-sessions and put them in a class library so they become internal resources for reviewing and micro-learning.
- **Digital Gallery Walks:** Interactive displays where students leave voice/text comments. No tech? Students can walk past each project, gallery style, and leave sticky note comments with feedback, questions, or ratings.
- **Question Trails:** Students create a project FAQ with answers from the gallery walk. Can be in text, with video, or with audio notes.
- **Virtual Showcases:** Live-streamed presentations to partner classrooms, lower grades, or even record the video to share with parents.

Pro Tip: Use AI tools to help students prepare concise talking points and identify key areas that might need clarification for their audience.

Teaching Real-World Communication

James Humes once said communication is the language of leadership, and isn't that what we're hoping for - that our students will become tomorrow's leaders (and then thank us for it in their autobiography)? Building marketable communication

291

skills turns PEACE projects into workplace training, giving them a leg up in life after graduation. Here's how to put this language of leadership into practice:

Social Media Marketing Madness

Wait a sec. When you read this header, did you think that we were unleashing our students on TikTok and the Gram? Um… no. This is not about having students sign up for every social media platform and then having to follow and filter for every student. This is about learning the mechanics of crafting social media in a way that can concisely share ideas and also be engaging. Students may never go viral or be trending, but what they do learn is the process of marketing their ideas.

1. **Social Media Snapshots**
   - **Instagram:** Share a captivating image with copy & hashtags
   - **Tweet (so vintage):** Summarize your project information in 280 characters or less
   - **TikTok:** Create a 60-second explanation of your project or findings
2. **Digital Documentation**
   - **Blog posts:** Detail your project journey along the way or share final thoughts

- **Podcasts:** Present your findings interview-style
- **Infographics:** Design visual data stories

3. **Professional Communication**

- **Email:** Draft summaries to experts, peers, or even principals!
- **Newsletter articles:** Craft a summary that captures interest and teaches
- **Grant proposal:** Write a summary proposal for funding further enquiry

Teacher Tool: Create a "Communication Choice Board" where students select different ways to share their work, tailored to both audience and purpose. You can organize the board with audience categories in the columns (such as friend, teacher, industry professional) and project types in the rows, offering a wide variety of communication options for each choice!

Reflection Routines That Actually Work

Because "What did you learn?" is possibly the least effective reflection question ever.

Daily Reflection Prompts

Looking for a new bell ringer that supports the PEACE process? Try these:

Start of Class:

- "Today I hope to discover..."
- "I'm still confused about..."
- "I want to explore..."

End of Class:

- "Today's biggest surprise was..."
- "I want to investigate further..."
- "I changed my thinking about..."

Project Milestone Reflections

As students work through the PEACE process, they sometimes make huge leaps in understanding or small steps towards their completed creation. Either way, we want to incorporate ways to reflect upon growth.

- **Question Review Circles:** Students write their original question from the Enquire phase, post them around the room, and then physically stand by their original questions. Next, students move to questions they can now answer. It might not have been their original question, but that doesn't mean they didn't learn the answer! Once there, they can discuss their ideas with a peer or write a reflection.

- **Learning Journey Maps:** Students create visual representation of their path, marking "aha moments" along the way and identifying turning points.
- **Expert Interviews:** Students interview each other as "experts" and document new understandings while also generating next-level questions.

The Power of Real Audiences

There is some serious power in sharing information with real audiences. Here's what happens when we move beyond classroom walls:

- Students invest more deeply
- Quality naturally improves
- Real-world skills develop
- Authentic feedback flows
- Community connections grow

That means including internal sharing as part of the Engage phase, like with classroom presentations, peer feedback sessions, and reflection journals. Then, take it out of your classroom and share with the school community, with partner classroom teaching, school newsletter, hallway exhibitions, and classroom websites that parents can access. Finally, consider going beyond school walls with community presentations, connecting with experts, and digital sharing.

Troubleshooting Common Challenges

Is all this going to be a breeze? Change never is. If you plan wisely, you can avoid falling prey to common challenges. Here are a few of those challenges with ideas on overcoming them:

Time Management

You're not the only one who started a 45-minute project that lasted three weeks. Instead, stay on top of time and still manage to get to all of your instructional moments with these approaches:

- Use station rotation for sharing instead of having your whole class listen to every presentation. As students come to your teacher station, where you have your assessment rubric ready, they present it to you and a small group while the rest of your students are working on other parts of the PEACE process or on more traditional classroom activities.

- Implement parallel presentation groups by having all your students present at once, just in small group settings. Your task will be to float around the room gathering assessment information simultaneously.

- Leverage digital platforms for asynchronous engagement by having students digitally deliver their presentation. Then assign students to small groups tasked with offering feedback and writing up a short review with further enquiry questions on the projects of their fellow group members.

Technology Access

The 1:1 revolution is still unrealized in many K12 classrooms in this country. The idea of a classroom set of digital devices for every student is a decades-long unrealized dream. Not every teacher is in an instructional setting where they have ideal technology access. Here are some ideas:

- **Device-Free Alternatives:** Explore traditional methods like role-playing, debates, or creative writing as Create phase options that don't rely on technology.

- **Sharing Pairs:** Instead of whole-class presentations, pair students up to make better use of a single computer station you may have in your classroom.
- **Hybrid Approaches:** Combine traditional and digital tools you can use in the library or computer lab. For example, use digital tools for research and collaboration, but then have students present their findings in a traditional format like a poster or slideshow back in the classroom.

Audience Building

Having an audience to engage with is a huge part of this last phase. If you've ever been to a cringy student play or disappointing science fair, you know people aren't exactly busting down your door to see what you're up to. If you're feeling a little light on the eyeballs, consider these strategies:

- **Buddy Classes:** Partner with a neighboring classroom or school to collaborate on projects. You may even try a partner class overseas!
- **Expert Networks:** Connect with experts in your field. They may not be able to give you a ton of time, but even a little can help.

- **College Experts:** Sometimes it's not about finding an expert per se, it's about finding someone who's further along in the learning journey than your students are. College students may mentor your class (and even earn themselves credit).
- **Community Partnerships:** Connect with local organizations and businesses to see if they are interested in collaboration. Some senior centers jump at any opportunity to support local school children.

Remember, every challenge is an opportunity to learn and grow. By addressing these common pitfalls head-on, you can leap over the stumbling blocks that slow others down.

In a Nutshell: The Engage Phase

Think of the *Engage* phase as your project's victory lap - but one where everyone gets to join in the celebration. It's where individual discoveries become community knowledge, and where students transform from learners into teachers.

The magic happens when we:

- Create authentic audiences
- Teach real-world communication
- Build in meaningful reflection
- Promote ongoing curiosity

Remember: The goal isn't just to share what students learned - it's to create ripples of impact that extend far beyond your classroom walls. When a student can teach something they've discovered, defend their findings with evidence, and inspire others to learn more, you know they have met the learning objective and blown it out of the water! When students successfully teach information and share their learning, that's when you know you've nailed the Engage phase.

And here's the beautiful part: Those ripples often come back as waves of new questions, leading us right back to Provoke, and starting our PEACE journey all over again.

Next up: Putting It All Together - Where we'll explore how these five phases create a seamless cycle of enquiry-based learning... and your plan to put them into action.

Chapter 5

Putting The PEACE Framework in Action

Remember when we started this journey? We talked about that moment when you first heard about AI in education and thought, "Great, one more thing to add to my plate." But here's what we've discovered: The PEACE Framework isn't about adding more - it's about making what we already do more powerful.

The Steps of PEACE

Step 1: Provoke 👆

Where curiosity gets its spark...

Ready to light the learning fire? This isn't about lecturing; it's about awakening student wonder. We are looking for 'ruler moments' where students are

301

buzzing with questions before you've even started teaching.

How? Through strategic provocations:

- Connect to **Prior Knowledge**: Quick writes, verbal brainstorms, media discussions.
- Spark **Curiosity**: Open-ended questions that challenge assumptions.
- Welcome **Wonder**: Let students consider questions like "What if..."

Step 2: Enquire 🦫

Turning wonderings into wonder-full questions.

Remember when your best research started with a simple "I wonder..."? Here's where students learn to transform vague curiosities into laser-focused enquiries.

Strategies include:

- The **BID Routine**: Brainstorm wildly, identify promising questions, and draft with precision using elements. Then **ReBID** to review, evaluate, and restart the iterative questioning process.

- Peer Collaboration: Because two curious minds are better than one.
- The **SHARP Filter**: Ensuring questions are Specific, High-level, Actionable, Relevant, and Powerful.

Step 3: Analyze 🔍

Information isn't knowledge... It's raw material.

Welcome to the research phase with an AI twist. Students aren't just gathering information; they're transforming it into meaningful knowledge.

Key moves:

- The **SAFE Check:** Evaluate information like seasoned fact-checkers by examining the source, accuracy, fairness, and evidence.
- The **CLEAR Check:** Refine AI interactions by ensuring responses are Complete, Logical, and well-Explained, and Adjusting the output or Rewording queries to improve results.
- **Time Management:** Break tasks into manageable chunks to keep the research process focused and productive.

Step 4: Create 🌑

Learning becomes tangible.

This is where understanding gets a creative passport. Students aren't just absorbing information; they're *demonstrating their learning* in ways that minimize cheating, support meaningful assessment, and that students actually enjoy!

Approach:

- **Meaningful Project Design:** Create projects that connect learning to real-world applications and student interests.
- **Authentic Expression:** Allow students to demonstrate their learning in diverse, creative ways.
- **Flexible Assessment Strategies:** Use assessments that give students multiple ways to showcase their understanding.

Step 5: Engage 🌑

Learning escapes the classroom.

The grand finale... where student work finds its real-world purpose.

Possibilities include:

- **Community Presentations:** Share student work with local audiences to make an impact.
- **Local Competitions:** Encourage students to apply their learning in competitive, real-world scenarios.
- **Online Publications:** Publish student work on digital platforms to reach a wider audience.
- **Real-World Problem Solving:** Tackle real-life issues through student-driven projects.

Because true learning doesn't have walls... **It has impact.**

Final Thought: The **PEACE Framework** isn't just a teaching method. It's a shift in how we think about learning... It's curious, connected, and completely alive. If you want to transform your classroom, this is your blueprint.

But I get it. Right now you might be thinking, "This all sounds great in theory, but my classroom doesn't exist in theory - it exists in fifth period right after lunch." Perhaps these bite-sized pieces won't give you indigestion.

Your First 30 Days with PEACE

Let's break this down into manageable steps, because nobody transforms their teaching overnight.

Week 1: Dipping Your Toes In

- **Monday:** Pick your guinea pig lesson (preferably one you could teach in your sleep), because starting with familiar content lets you focus on the new process, not the material.
- **Tuesday-Wednesday:** Brainstorm one truly juicy question for your Provoke phase, the kind that makes kids' heads pop up from their phones - you know the one. (Phase: Provoke - Spark Curiosity)
- **Thursday-Friday:** Test drive a simple AI tool like Claude, Gemini, or ChatGPT. Start with something basic, like generating discussion questions, checking student work, or getting feedback on your juicy thought-provoking question from yesterday.
- **Weekend Reflection:** What worked? What didn't? What surprised you? Keep a quick note on your phone - nothing formal, just real observations.

Week 2: Finding Your Rhythm

- **Monday-Tuesday:** Provoke your student with your curiosity question and have them Think-Pair-Share to generate questions. *(Phase: Provoke - Brainstorm)*
- **Wednesday-Thursday:** Practice crafting better questions with students. Show them how to turn "When was the War of 1812?" into "Why did they even call it that?" *(Phase: Enquire - Model)*
- **Friday:** Document one success story. Catch a student asking a killer question? Write it down! That's gold.
- **Weekend Planning:** Map out next week's adventure, but keep it flexible - teaching is jazz, not classical music.

Week 3: Leveling Up

- **Monday-Tuesday:** Introduce basic AI-assisted research techniques. Show students how to fact-check AI responses and look for the truth. No tech? No problem! You can share the process with students on your interactive whiteboard. *(Phase: Analyze - SAFE & CLEAR Check)*
- **Wednesday-Thursday:** Build in structured reflection points. Quick writes, exit tickets, or my

personal favorite: "What confused you today?" *(Phase: Create - Assessments)*

- **Friday:** Share wins with a colleague, because teaching in isolation is like dancing alone - possible but way less fun.
- **Weekend Assessment:** Review student responses. Look for patterns, surprises, and those magical "aha" moments.

Week 4: Putting It All Together

- **Monday-Tuesday:** Integrate multiple PEACE elements in one lesson. It's like adding ingredients to a soup - start tasting as you go and adapt.
- **Wednesday-Thursday:** Try more advanced AI applications. Maybe have AI play devil's advocate in a class debate?
- **Friday:** Celebrate progress with students, because sometimes we all need to stop and say, "Hey, look how far we've come!"

Remember: It's okay if it's messy at first. Every master teacher started as a beginner.

***

The PEACE Quick-Start Guide

You know that feeling when you're trying a new recipe and the instructions say "cook until done"? Helpful, right? Let's be more specific about implementing PEACE. Think of this as your "Choose Your Own Adventure" guide to transforming your teaching - minus the dead ends and dragon encounters.

Choose Your PEACE Adventure

1. **The Cautious Approach**
   - Start with traditional lessons
   - Add one PEACE element at a time
   - Gradually integrate AI tools
2. **The Deep Dive**
   - Select one unit to fully transform
   - Implement all PEACE phases
   - Use AI throughout the process
3. **The Hybrid Method**
   - Keep some traditional lessons
   - Transform others completely
   - Build confidence gradually

The beauty of PEACE is that there's no wrong way to start - there's just your way. Whether you're tiptoeing in or diving headfirst, you're moving in the right direction.

***

Making PEACE Work in Your Reality

Time Management Strategies

If there's one thing we all wish we had more of, it's time. Wouldn't it be great if we could control time like Evie in "Out of This World" and start and stop it with a simple touch? Since we aren't half alien, we'd better put more human-like strategies into place. Try these:

- **Use AI for routine tasks:** Let AI handle the basic stuff like generating practice problems or providing initial feedback on drafts. Remember, AI is your teaching assistant who never takes a sick day, but they won't do the work unless you ask for it. Play around and find the best ways for AI to make *your* life easier. Also work on building a personal swipe file of prompts that make AI do what you want it to.

- **Build question banks:** Create and save great thought-provoking questions - they're like lesson plan money in the bank. One good question can launch a dozen great discussions.

- **Create reusable protocols:** Routines are everything. The more you can create reusable chains of behaviors, the better you'll manage all your students working on different tasks at the same time. Design your Provoke and Enquiry phases in a structured way that works across multiple units, that way you can insert the fresh topic, and students understand the expectations.

- **Choose flexible assessment tools:** We covered a lot of different ways you can assess students learning at multiple checkpoints throughout the PEACE process. Find the ones that make sense to you and your students. Prep them ahead of time, get comfortable with them, and have them available.

Resource Management Strategies

Look, we've all been there - staring at our wish list of classroom resources and a bank account that looks like a bad joke. But here's the good news: Implementing PEACE doesn't require a tech startup's budget.

Starting Simple: The "Use What You've Got" Approach

Your Best Friend: Free AI Tools! There are a lot of paid teacher AI platforms out there, but you can absolutely start with the free versions of tools like:

- Claude (perfect for generating thoughtful feedback)
- ChatGPT (great for creating practice problems)
- Google Gemini (excellent for brainstorming)

Pro Tip: Create a shared document with your department to track which free tools work best for different tasks. It's like a recipe book, but for AI tools. Then collectively add prompts that work well so you can build a better swipe file.

Class Tech: The "It Is What It Is" Approach

Remember that cart of slightly outdated Chromebooks in the back of your room or the tablets the librarian keeps in the closet? They're about to become your secret weapon. Here's how:

- Set up browser bookmarks for your go-to AI tools
- Create QR codes for quick access to digital resources
- Design activities that work with small groups and peer pairs when you're short on devices

Student Devices: The "Elephant in the Room" Approach

Let's be real - most of your students are carrying around more computing power than NASA used to reach the moon. Instead of fighting it:

- Create clear protocols for appropriate device use
- Design activities that work with or without devices
- Have backup plans for students without access

Digital Organization: The "Marie Kondo" Approach

- Create a simple folder structure for saving great AI prompts
- Build a digital repository of successful lessons
- Keep a running document of what worked (and what spectacularly didn't)

313

When it comes to resource management, start simple and use what you've got. It is what it is when it comes to tech access, so design for available devices. If your students have smartphones, come up with a system to manage them (and use them for good and not evil). Finally, have a system to keep organized! There's nothing worse than coming up with a great prompt or a wonderful lesson, and then not being able to find it when you need it.

Classroom Management Strategies

You know that moment when the lesson plan meets reality. Sometimes it's beautiful, like fireworks. Sometimes it's just explosive. And sometimes it's a dud. Let's talk about managing the beautiful mess that is AI-enhanced learning.

Setting Up: Those Foundation First Days Matter!

- Establish clear signals for switching between activities
- Practice AI tool protocols before you need them
- Create simple troubleshooting guides students can follow

Flexible Grouping: Cue Up the Band!

Think of your classroom like a jazz ensemble - sometimes solo work shines, sometimes you need the whole band:

- Design roles that rotate naturally (AI Fact Checker, Question Master, Synthesis Specialist)
- Create clear transition signals between group configurations
- Build in quick reflection points between changes

Keeping It All Together: Checkpoint Charlie

Nobody likes the "deer in headlights" moment when you ask, "Any questions?" Instead:

- Use digital check-in forms or visual hand gestures (thumbs up, thumbs down, or in the middle) for quick temperature checks.

- Create visual progress markers for longer projects and put this in your classroom where students can easily access it and move themselves along as they work. This way you can track everyone at once.

- Build in peer review stations for work-in-progress sharing so students have natural deadlines to keep them working. It's also a time to follow up with students that you think are struggling to keep up and provide them with extra support.

When Things Go Sideways (Because They Will)

Have ready-to-go backup plans that might include:

- No-tech versions of your activities or at least easy ways to adapt them offline.
- Realignment moments where you can rewind and reframe the directions, with more modeling built in and some solo reflection.
- Analog alternatives that still hit your learning targets, so don't burn everything in that ol' filing cabinet just yet!

***

The Power of PEACE: Bringing It All Together

Think of PEACE as less like a rigid framework and more like your favorite recipe - you know the basic ingredients, but sometimes you add a little extra spice based on what your students need that day.

Each phase isn't just a checkbox - it's an opportunity:

Provoke: Remember that student who never raises their hand? Watch them light up when you ask the right provocative question. It's about creating those lean-in "ruler moments" when suddenly everyone's paying attention because they can't help themselves. *The goal:* Spark curiosity, activate prior knowledge, and set purpose for learning.

Enquire: Wonder runs wild and when students start asking questions you never thought of, you'll know you're on the right track. It's like watching them unlock doors they didn't even know existed. *The goal:* Develop question quality, build research skills, and focus investigation.

Analyze: Critical thinking isn't just a buzzword here… it's watching students learn to navigate a world where information is unlimited but wisdom is earned. Guide them to become information skeptics, not just information consumers. *The goal:* Enhance critical thinking, improve data literacy, and strengthen evaluation skills.

Create: This is where students shock you with their brilliance. Give them the tools, step back, and watch them build things you never imagined. Sometimes the best teaching is knowing when to get out of their way. *The goal:* Demonstrate understanding, encourage innovation, and apply learning.

Engage: Learning in isolation is like trying to clap with one hand. Create opportunities for students to share their discoveries, challenge each other's thinking, and build on each other's ideas. *The goal:* Share discoveries, build communication skills, and create community impact.

When you go all in on the PEACE Framework, you're not just following a set of steps - you're creating an inclusive environment that pushes every student to become an active learner and critical thinker. By striking the right balance between this structured framework and flexible teaching, you build a space where students can explore, question, and create in ways that resonate with their unique experiences and interests, all while reinforcing the academic standards they're expected to meet. Ultimately, this approach has the potential to reframe educational practices, integrate 21st-century technology and skills, and ignite a passion for lifelong learning within your students.

Why This Matters Now

In a world where AI is rapidly changing how we work and learn, the PEACE Framework isn't just another teaching method - it's a bridge to the future. As artificial intelligence continues to shape every aspect of our lives, students need a strong foundation in AI literacy. This tech isn't going anywhere! Beyond knowing how to use AI, they must understand its capabilities, limitations, and the ethical considerations surrounding it. This literacy is critical as the skills required in the workplace evolve to emphasize adaptability, digital proficiency, and problem solving. Plus, with AI shifting the landscape of information access, students need more than a passive relationship with knowledge; they need the ability to critically engage with and apply what they learn.

At the heart of the PEACE Framework is enquiry, and in today's rapidly evolving world, those skills are essential. Unlike the traditional emphasis on memorization, critical thinking and problem solving are quickly taking precedence. Students must learn to navigate complex questions, assess information critically, and apply their understanding in dynamic contexts. In an environment where change is constant, adaptability is no longer just an asset - it's a necessity. By focusing on enquiry, the PEACE Framework targets these skills and builds capacity amidst uncertainty.

And let's not forget how teaching has changed in our post-pandemic classrooms. We desperately need something to make learning feel relevant and authentic for students who are finding it harder and harder to engage in learning. The PEACE Framework does that. By grounding lessons in real-world contexts, it builds connections that go beyond textbooks, resonating with students on a personal level. This approach boosts engagement, as students see the practical implications of what they learn, making the classroom experience feel less abstract and more meaningful. As students tackle real-world problems, they naturally develop skills that transfer to life beyond school. With the

PEACE Framework, learning becomes a dynamic, impactful experience that prepares students not only to succeed in today's world but to help shape tomorrow's.

Your Next Steps

You have the blueprint, but it's up to you what happens next. Commit to this process:

Start Tomorrow: Choose one lesson, add one PEACE element, and try one AI tool.

Build Over Time: Document what works, share with colleagues, and celebrate small wins.

Keep Growing: Join learning communities, stay current with AI, and refine your practice.

You can do this! You're capable of anything if you put your mind to it.

Progress Not Perfection

The PEACE Framework isn't about perfection - it's about progress. Every time you:

- Let students drive enquiry

- Integrate AI thoughtfully
- Create space for creativity
- Build in reflection time
- Share learning widely

You're not just teaching content - you're preparing students for a future where the ability to ask good questions matters more than knowing all the answers.

Your PEACE Promise

Make this commitment to yourself and your students:

- I will start small but think big.
- I will embrace the messiness of learning.
- I will grow alongside my students.
- I will celebrate progress over perfection.
- I will keep what works and refine what doesn't.

Because at its heart, the PEACE Framework is about more than just teaching - it's about transforming how we learn together in an AI-enhanced world.

Are you fired up? Now it's time to think about what's happening outside of your classroom.

Chapter 6

Navigating the Challenges: Making PEACE with AI

I remember one of the first times I tried using AI during a lesson. There I was, standing in front of twenty-eight pairs of eyes, trying to explain how we were going to use this "really cool new tool" when suddenly... the internet went down.

Have you been there? That moment when your carefully planned tech-enhanced lesson meets reality?

Let's be honest - bringing AI into your classroom isn't just about following a framework or implementing new strategies. It's about navigating real concerns, handling legitimate fears, and figuring out how to make this work in the beautiful mess that is everyday teaching.

*****

The Questions We're All Thinking (But Maybe Afraid to Ask)

"So... *Do I really need to learn AI?*"

This question hangs in the staff room, carrying more anxiety than an unannounced Monday morning classroom observation by your evaluator. Do I think AI will replace teachers? *I hope so.* Wait, what? As a nation, we are facing huge teacher shortages. I can imagine a world where teachers who make worksheets and ask Google-able questions are replaced with AI in some way just to cover all our classes.

On the other hand, AI can never be a complete replacement for teachers, especially for those that have transformed their teaching and lead dynamic learning. That means you don't *need* to learn AI, but don't you *want* to?

Think of it like your first smartphone... it was once intimidating, now it's indispensable. Learning AI isn't about job preservation; it's about professional transformation. Teachers bring something to the classroom that AI never will - the ability to look a student in the eye and *know* they didn't eat breakfast, to sense when the usually chatty kid in the back row is suspiciously quiet. AI can generate content, but *you* make learning come alive. Your superpower isn't algorithmic precision; it's *humanity*. So start small. Experiment with one AI tool this month. Ask curious questions. Treat it like a collaborative buddy, not a threat. Because at the end of the day, you're not just delivering information… you're building *humans*.

"How do I build a support network when I barely have time to eat lunch?"

We are all in the same boat here! AI in education is something we're all figuring out together. Take classes, watch YouTube videos, chat with your favorite LLM - be open to growing and building a network when and where it's easiest.

Remember when you first started teaching and that veteran teacher down the hall became your unofficial mentor? Finding your AI support crew isn't that different. Start with one person - maybe that tech-savvy teacher who's always trying new things. Share one success and one epic fail over coffee. Before you know it, you've got your first AI teaching buddy.

Pro tip: Create a shared digital space (even a simple Google Doc works) where you can dump quick wins, funny fails, and "Hey, did you try this?" moments.

It's important to stay connected with other educators and seek opportunities for ongoing learning. Connect with me on LinkedIn at www.peaceframework.com/linkedin

---

"How do I assess anything when AI can basically do everything?"

Oh, this is a fun one! Remember when we thought calculators would make math irrelevant? (Spoiler alert: They didn't.) During every wave of technology advancement, there's been someone yelling "The sky is falling!" Instead of seeing AI as the enemy of assessment, think of it as a way to level up your assessment game.

Consider these real-world examples:

In Math: Instead of just asking students to solve word problems, have them use AI to generate three different solution strategies. Then students analyze which approach is most efficient, explain why, and create their own unique solution method that improves upon the AI's suggestions. *Bonus: Watch them discover that sometimes the "efficient" way isn't always the most understandable way. #NewMath*

In Science (Middle School): Rather than having students simply write up their lab results, have them use AI to generate three different hypotheses about why their plant growth experiment had unexpected results. Students then design additional experiments to test each AI-generated hypothesis, explaining which they think is most likely and why.

In Social Studies: Instead of asking students to write a traditional essay about the causes of the American Revolution, have them use AI to generate perspectives from three different historical figures. Students then analyze these viewpoints, identify potential biases, and construct their own evidence-based argument about which perspective they find most compelling.

In 2nd Grade: Instead of just asking students to write about their favorite animal, have them use AI (with teacher guidance) to generate three different interesting facts about that animal. Students then draw pictures showing why they think each fact is true or false based on what they already know, and share their reasoning with the class. *You haven't lived until you've heard a 7-year-old explain why they don't believe an AI's claim about penguin sleeping habits.*

In Music: Rather than just having students compose a melody, have them use AI to generate three variations of their original tune. Students analyze how each variation changes the mood or feeling of the piece, choose their favorite elements from each version, and create a final composition that incorporates their preferred elements.

The key in all these examples is moving from "show me what you know" to "show me how you think." When students engage with AI-generated content critically, they're not just learning the subject matter - they're developing analytical skills that no AI can replicate.

Pro tip: Keep a running collection of your best AI-enhanced assessment ideas. That worksheet you just transformed might be exactly what your colleague needs next semester.

"What about privacy? I can barely keep track of who's allowed to be in class photos!"

First, breathe. A lot of these decisions will come from your district (*thank goodness*), but there's still plenty you can do:

- Use class accounts instead of individual student accounts
- Teach students to anonymize their data (no names, no personal details)
- Create clear guidelines about what can and can't be shared with AI
- When in doubt, err on the side of caution

"How do I share student work without compromising privacy?"

Remember classroom bulletin boards? Think of digital sharing the same way:

- Create a class showcase space (physical or digital)
- Use pseudonyms or student numbers
- Focus on the work, not the creator
- Let students choose what they're comfortable sharing

"How do I develop AI literacy when I'm still figuring it out myself?"

Confession time: Nobody has this figured out completely. The best teachers I know are the ones who model being a learner. When something goes wrong (and it will), say "Wow, I didn't expect that! Let's figure out why together."

Make your classroom a place where:

- Questions are celebrated
- Mistakes are learning opportunities
- Everyone (including you) is learning
- "I don't know, let's find out" is a perfectly acceptable answer

"How do I build a classroom culture that embraces AI without becoming dependent on it?"

Think of AI like a sous chef in a cooking class - incredibly helpful for prep work and technique, but the students still need to understand the fundamentals of cooking and develop their own culinary instincts. Create spaces where:

- Students know when to use AI and when to rely on their own thinking

- Technology amplifies student creativity rather than replacing it
- Critical thinking and problem solving drive every interaction
- The focus stays on learning, not the tools we use to learn

Classroom Vignette: Embracing AI in Professional Growth

During a department meeting at Lincoln High School, veteran English teacher Mrs. Thompson voices what many are thinking: "But won't AI just write their essays for them?"

The Journey

Initial Concerns:

- Student cheating
- Loss of authentic writing
- Grading fairness
- Job security

Collaborative Solution-Finding

Department Chair Dr. Martinez addresses her staff at an in-service day.

"I understand that the idea of integrating AI into our classrooms can feel daunting. Many of you are probably feeling a mix of emotions right now... curiosity, excitement, perhaps even some anxiety that AI is going to take your jobs. But trust me, no robot can ever replace the relationships you have with students! Besides, who else would be there to fix the copy machine when it jams? Just remember this: We're all in this together. None of us are AI experts (yet!), but we can learn from each other's experiences and support each other as we make AI part of our team."

She addresses several agenda items, including:

- Sharing specific teacher concerns
- Exploring AI tools together
- Accessing quality online professional development training
- Testing instructional strategies in small groups
- Celebrating success stories

Implementation in Action

Mrs. Thompson's Evolution:

1. Week 1: Uses AI to generate personalized writing prompts
2. Week 3: Discovers AI can help create scaffolded outlines for struggling writers after attending online training
3. Week 6: Develops AI-assisted peer review guidelines
4. Week 8: Students use AI to explore multiple perspectives on their thesis statements

Key Breakthrough

"I realized AI wasn't replacing my teaching - it was giving me more time to actually connect with students about their writing. Now I spend less time creating basic materials and more time having meaningful conversations about their ideas."

Department Impact

- Increased student engagement in writing
- More individualized feedback
- Reduced teacher planning time
- Improved student-teacher relationships

Teacher Reflection

"My fear of AI came from not understanding its role in my classroom. Once I saw it as a tool to enhance rather than replace human connection, everything shifted. Now my students are more excited about writing than ever, and I'm a better teacher for being open to change."

***

Remember This

You're not just teaching with AI; you're preparing students for a world where human skills like empathy, creativity, and ethical decision-making matter more than ever. The technology will change (probably before you finish reading this sentence), but good teaching will always be about connecting with students and helping them grow.

Your Permission Slip

Dear Amazing Teacher,

You have permission to:

- Start small (really small - like "one-activity-in-one-class" small).
- Make spectacular mistakes (they make the best stories).
- Change your mind (what worked Tuesday might bomb on Wednesday).
- Take breaks from AI (sometimes old school is the best school).

- Celebrate tiny wins (did one student ask a better question? #VICTORY).
- Put relationships first (always and forever).

Next Steps

As we move into exploring specific activities, remember: These are starting points, not prescriptions. Take what works, modify what doesn't, and make it your own. Because that's what great teachers do - we beg, borrow, steal, and adapt until we find what works for our students.

And isn't that what teaching has always been about?

Chapter 7

AI-Powered Activities for Kids

Incorporating large language models (LLMs) into classroom instruction offers practical opportunities to enhance essential skills, such as executive functioning, prompt engineering, and critical thinking. This section provides activity ideas that help teachers effectively integrate AI tools into their curricula, supporting students in developing these critical skills.

These activities focus on specific, actionable strategies for using AI to foster organization, problem solving, and creative thinking in students. By utilizing LLMs, educators can provide personalized learning experiences tailored to individual student needs, making it easier to address diverse learning styles and abilities. While they may not fit neatly into an enquiry-based learning model or the PEACE Framework, you

337

can incorporate key elements of questioning and collaboration throughout.

Each lesson plan includes clear objectives, practical activities, and suggestions for AI integration, enabling teachers to facilitate meaningful interactions between students and AI. This approach not only supports academic achievement but also prepares students for future challenges in a technology-driven world.

Prompt Engineering and AI Interactions

These activities emphasize developing students' ability to effectively use and interact with AI tools, specifically focusing on prompt engineering. The goal is to help students understand how to craft clear, specific prompts that will generate desired responses.

Say What, Alexa?

This activity challenges students to reverse-engineer prompts by analyzing AI responses, encouraging them to think about how different prompts yield different results.

Procedure:

1. **Generate Prompts and Responses:** Create a list of prompts for Alexa and the responses those prompts would generate.
2. **Provide Responses Only:** Give students only the responses, without the prompts.
3. **Pair Up:** Have students work in pairs to guess the verbal prompt that might have led to each response.
4. **Group Comparisons:** Combine two pairs of students into groups of four. Each group will compare their guessed prompts and choose the best one for each response.
5. **Reveal Prompts:** The teacher will share the actual prompts. Students will then evaluate their guesses.

Target Skills:

- Communication Skills
- Prompt Engineering

- Problem-Solving Skills
- Critical-Thinking Skills

Variations:

- **Try It Yourself!** Have students partner to predict the prompt, then actually prompt Alexa to see if they get the expected response.

- **Another GPT:** Students can compare several LLM models to see which provides the response that is closest to the target.

Reverse Image Prompt Engineering

Students aim to recreate an AI-generated image by developing text prompts, highlighting the connection between language and visual output in AI.

Procedure:

1. **Provide an AI-Generated Image:** Start by showing students an AI-generated image.

2. **Observe and Describe:** Ask students to carefully observe the image and write down 3-5 keywords that describe it.

3. **Create a Text Prompt:** Have students develop a text prompt that they believe could generate a similar image.

4. **Generate an Image:** Students will use an AI image model to create an image based on their written prompt.

5. **Compare Outcomes:** Students pair up or form small groups to compare their generated images with the original. Then, score their images based on how closely they resemble the original.

6. **Reveal the Original Prompt:** Finally, the teacher will share the original AI prompt. Students can then compare it to their own prompts.

Target Skills:

- Observation Skills
- Prompt Engineering
- Critical-Thinking Skills

Variations:
- **You Write the Prompt!:** Before starting, have students write the prompts you will use to generate the image. This makes the lesson more engaging. Remind them to include:
  ○ The medium (e.g., photo, painting)
  ○ The subject (e.g., a cat, person, car)
  ○ The context (e.g., in a park, on the beach)
- **Best Outta Three?:** Have students create three variations of their prompts to achieve their target results. They can then rank these prompts based on which one produced the best outcome.
- **Wait, Lemme Try Again:** Require students to create an image with their first prompt, then rewrite that prompt to improve their result. They will bring their revised prompt and image to their small groups for ranking.
- **Gallery Walk:** Instead of small group discussions, have students display their images in a gallery walk. All students can then score the images based on a preset rubric.

Can You Do It Better?

Students iteratively refine prompts to achieve better AI responses, learning the importance of precision and clarity in their instructions.

Procedure:

1. **Provide a Prompt:** Present students with a prompt. You can project it on an interactive whiteboard or distribute it on a worksheet.
2. **Iterate the Prompt:** Ask students to improve the prompt to get a better result. They should revise the prompt three times, ending up with three improved versions.
3. **Peer Evaluation:** Have students pair up with a partner to evaluate each other's updated prompts.
4. **Final Revision:** After discussing, each student should select their best version and make one more update based on their partner's feedback.
5. **Access the LLM:** Students will use an AI large language model (LLM) to enter their four prompts: the three drafts and the final version.
6. **Score the Responses:** Each student will score the responses based on a provided rubric to determine which prompt yielded the best result.

7. **Final Iteration:** Students will create a new prompt based on their drafts and the AI responses. They will write this down and submit it for grading.

Target Skills:

- Communication Skills
- Critical Thinking
- Prompt Engineering
- Problem-Solving Skills

Variations:

- **More Is Better?:** Take it a step further by having students enter their final prompt into three different LLM models. They can compare results to see how the dataset affects the responses.

- **Whatta Ya Think?:** After scoring, students can meet in small groups to discuss what worked, what didn't, and what changes they made. This reflection helps improve their questioning skills, emphasizing the importance of iteration.

Iterated

Students use AI to generate variations of a prompt, exploring different contexts and directions within a topic, further developing their prompt-engineering skills.

Procedure:

1. **Choose a Line of Enquiry:** Select a specific topic for students to explore. This could be a particular book, event, or mathematical concept.
2. **Initial AI Prompt:** Instruct students to type the following prompt into the AI: "I am going to give you a prompt. Return 5 variations of that prompt that give a different context or direction to the original question. Start by asking me for my question."
3. **Ask a Question:** Students will then provide their own question based on the guidelines you've set.
4. **Evaluate AI Responses:** After the AI generates variations, students will evaluate the questions produced and choose the one that interests them the most.
5. **Reinforce Iteration:** Discuss how this process reinforces the concept of iterations in prompt

engineering, emphasizing that refining prompts leads to more insightful responses.

Target Skills:

- Critical Thinking
- Prompt Engineering

Variations

- **Follow the Question:** Although this is focused on iterating prompts, you can have students follow their questions to logical conclusions. This is in alignment with the enquiry-based learning framework and supports a deeper understanding of the topic.

Mnemonic Song

Students use AI to generate song lyrics to help them remember mnemonic devices, key information, or even the plot of a book, which develops their creativity and critical thinking.

Procedure:

1. **Identify Target Skills:** Begin by selecting a specific set of skills that will help students access grade-level standards.
2. **Choose a Skill and Song:** Allow students to choose a skill they want to work on, and then let them pick a popular or favorite song.
3. **Adapt the Song:** Students will use an AI large language model (LLM) to generate lyrics that adapt their chosen song to focus on the target skill. For example, they might adapt "All the Single Ladies" to teach PEMDAS for order of operations.
4. **Share Creations:** Have students share their adapted songs with the class or in small groups.
5. **Evaluate Effectiveness:** Students will then evaluate whether the song helps them learn the skill, using a provided rubric.

Target Skills:

- Communication Skills
- Critical Thinking
- Prompt Engineering
- Creativity Skills

Variations:

- **Dub This!:** If students have access to audio AI tools, allow them to dub their songs (just remember to avoid using copyrighted music!).
- **Anchors Away!:** Have students create anchor posters or lyric sheets as part of their project. These can be displayed in the classroom for ongoing reference.
- **Cover Me:** Have AI help design the single's album cover. Use this as visual anchors in the classroom so students can recall the song when they need it.

Mnemonic Song Assignment Grading Rubric

Criteria	Excellent (4 pts.)	Proficient (3 pts.)	Basic (2 pts.)	Limited (1 pt.)
Creativity	Highly original and engaging adaptation of the song.	Shows creativity but somewhat predictable.	Some creativity, but lacks originality.	Lacks creativity or closely mirrors original song.
Clarity of Skill	Clearly conveys the target skill.	Mostly clear but could be sharper in parts.	Conveys the skill, but lacks clarity.	Fails to effectively convey the skill.
Communication	Clear, confident, and engaging delivery.	Clear, with some audience engagement.	Minimal clarity or engagement.	Delivery lacks clarity or engagement.

	Strong connections between song and skill.	Good connections, but could be deeper.	Weak connections between song and skill.	Vague or no connections made.
Critical Thinking				
Prompt Engineering	LLM-generated lyrics are clear and well-targeted.	Lyrics mostly align but lack precision.	Lyrics somewhat align but may lack coherence.	Poorly aligned lyrics, unclear or inconsistent.
Problem Solving	Excellent adaptation of song to fit target skill.	Good adaptation, though some areas need work.	Adaptation is incomplete or confusing.	Little to no effective adaptation.
Engagement	Actively engages classmates and seeks feedback.	Engages classmates, though less enthusiastic.	Minimal engagement with peers.	Does not engage or collaborate.

Total Score: **/ 28**

Additional Feedback:

Strengths:

Areas for Improvement:

350

Creativity and Storytelling

This set of activities focuses on using AI to build creative writing and storytelling skills. They encourage students to experiment with different narratives, characters, and scenarios while using AI as a brainstorming and content-creation tool.

My One Act Play

Students collaborate to write and perform short plays, using AI for inspiration with the story's beginning, combining creative writing with performance skills.

Procedure:

- **Choose a Title:** Decide on a title for the plays.
- **Generate the Story Beginning:** Use an LLM to create the opening of the story. Provide the AI with:
  - The topic
  - The number of characters
  - Character names
  - An unexpected element
  - *Bonus: Ask students to suggest character names to make the story more personal.*
- **Write the Play:** Have students pair up to write the rest of the play. They can use an LLM to help them generate ideas and dialogue.
- **Vote on Plays:** Once all plays are submitted, allow students to read each play and vote for their favorite.
- **Perform the Plays:** The class will perform the selected one-act play in small or large groups.

Target Skills:

- Communication Skills
- Teamwork and Collaboration
- Prompt Engineering
- Creative Writing
- Critical Thinking

Variations:

- **Bigger Small Groups:** Instead of pairs, form small groups of 4-6 students and include more characters in the story starter. This will make the plays more complex and engaging.
- **Student Scribes:** Allow students to provide details and use an LLM to generate the entire play. This will create diverse plays and provide a chance to compare outputs from different LLM models.

Martin Scorsese-d

Students develop storyboards and use AI to generate visuals, exploring visual storytelling and connecting narrative with imagery.

Procedure:

1. **Choose a Focus:** Select a specific standard, book, historical event, or mathematical concept for students to explore.
2. **Create a Story Outline:** Students will create an outline for their story based on the chosen focus. They can use an AI language model (LLM) to help generate ideas for the storyline.
3. **Develop a Storyboard:** Students will create a 10-frame storyboard. For each frame, they will write prompts needed to generate AI images that represent their scenes.
4. **Generate Images:** Using an image LLM, students will create the images needed for their storyboard frames.
5. **Optional Movie Creation:** Students can create a short movie using their 10 images, overlaying the audio of their storyline.

Target Skills:

- Creativity Skills
- Prompt Engineering

Variations:

- **Hey Partner!:** Students can work in small groups or pairs to create the final project.

My Cousin, Vinny

Students use AI to create humorous courtroom scenes, promoting creative writing within a specific format and encouraging imaginative scenarios.

Procedure:

1. **Choose Characters:** Allow students to select two unrelated characters, historical figures, or people.
2. **Select a Competition:** Have students decide on a task for the characters to compete in (e.g., "Who would win in a dance-off, Maya Angelou or Brett Favre?").
3. **Set Guidelines:** Students will establish guidelines for their scene using an AI language model (LLM). For example, they might specify: "The debate is a mock courtroom drama where each character presents their argument, followed by a humorous rebuttal."
4. **Evaluate and Revise:** Students will review the AI-generated results and can reprompt as needed to improve the outcome.
5. **Share and Rate:** Share the courtroom dramas with the class, allowing peers to read through them and rate the results or decide who won the competition.

Target Skills:

- Creative Writing Skills
- Prompt Engineering

Variations:

- **And the Winner Is:** Students can decide the winner from their personal exchange with AI to help the class decide on an overall champion.

Divergent

Students explore alternate storylines by changing a character's decision in a story, prompting AI to predict the new outcomes, encouraging creative thinking and problem solving.

Procedure:

1. **Choose a Character:** Have students select a character from a story.
2. **Identify a Key Moment:** Students will pick a moment in that character's story arc where they could have made a different choice.
3. **Predict Outcomes:** Using AI, students will predict how this change affects the story, the character, and other characters involved.
4. **Evaluate the Change:** Students will discuss whether the moment they chose to diverge from the original storyline made a significant difference in the new version.
5. **Partner Discussion:** Students will pair up to discuss the moment they picked, the alternate choice made, and the new outcome for that character.

Target Skills:

- Communication Skills
- Prompt Engineering
- Critical-Thinking Skills
- Problem-Solving Skills

Variations:

- **I Just Tagged You:** Have students use AI to create a picture of the character in their new ending. They should then write an Instagram, Snap, or other social media post to go along with it and tag a place, another character, or a mood that makes their post that much better.

CYOA

Students collaborate with AI to develop "choose your own adventure" stories, incorporating decision points that lead to various outcomes, encouraging interactive narrative design.

Procedure:

1. **Create an Interactive Story:** Have students use AI to create an interactive story where they make choices. They must prompt the AI to stop 5 times to ask deciding questions.

2. **Note Unchosen Options:** As students write the story with AI and make choices, they should note the options they didn't take.

3. **Create Accompanying Images:** At the end of the story writing process, students will create 3-5 images to illustrate their adventure.

4. **Share in Groups:** Students will meet in small groups or pairs to share their stories and explain the decisions they made along the way.

5. **Predict Alternate Outcomes:** While sharing, students can also predict what might have happened if they had made different choices.

Target Skills:

- Communication Skills
- Prompt Engineering
- Critical-Thinking Skills
- Problem-Solving Skills

Variations:

- **Play That One Back:** After the story is finished, students can create another story based on the options they didn't choose. Was this a better adventure?
- **That's Deep:** Students can create a presentation with touch interaction that includes all the decision moments and narrative for a true choose-your-own-adventure story.

Retake

In this activity, students will use an LLM to help write a creative retake on a modern movie, reimagining it with a book character or historical figure as the main character. This exercise will enhance their creativity, storytelling skills, and understanding of character development.

Procedure:

1. **Choose a Modern Movie:** Students will select a modern movie that they are familiar with. This could be anything, from action, to romance, to comedy.

2. **Select a Character:** After students have chosen their movie, reveal the book character or historical figure to serve as the new main character.

3. **Analyze the Original Movie:** Students will briefly analyze the original movie's plot, themes, and character dynamics. They should consider how the chosen character could alter the story and its outcomes.

4. **Use AI to Generate Ideas:** Students will brainstorm ideas for their retake with an LLM. They can prompt the AI with questions about plot

twists, character motivations, and dialogue that could fit within the new context.

5. **Write the Retake:** After brainstorming, students will write a short script or outline for their retake, making sure to highlight how the character's unique experiences and qualities influence the storyline.

6. **Present Their Retakes:** Students will share their creative retakes with the class, discussing their character choices and how they adapted the plot.

Target Skills:

- Creative Writing
- Character Analysis
- Storytelling Skills
- Critical Thinking

Variations:

- **Re-Takes Two:** Have students work in pairs or small groups to choose a movie and character together. They can then create a joint retake, allowing for collaboration on ideas and character interactions.

- **Genre Swap:** After stunning students with the switch characters, stun them again by changing the movie's genre, like creating a comedic retake of a thriller. This adds an extra layer of creativity (and fun).

- **I Journal:** Have students do more with this by writing journal entries from the perspective of the chosen character during key moments of the movie. It deepens their understanding of the character's motivations and emotions plus explores a different writing style.

- **Let's Get Visual:** Have students create a visual storyboard of their retake using an AI image generator to illustrate key scenes and character interactions.

Reality Show Remix

In this activity, students imagine a collection of different historical figures or fictional characters dropped into the plot of a reality show. This exercise will enhance their understanding of character traits, plot development, and creative storytelling.

Procedure:

1. **Select Historical Figures or Characters:** Generate a list of relevant historical figures and/or fictional characters. Have students randomly choose their cast.

2. **Choose a Reality Show Format:** Students will select a popular reality show format, such as a competition (e.g., "Survivor", "The Amazing Race"), a lifestyle show (e.g., "The Real World", "Coming from America"), or a talent show (e.g., "America's Got Talent", "America's Next Top Model").

3. **Analyze Character Traits:** Students will analyze the chosen historical figures or characters, noting their traits, motivations, and potential interactions with one another. They should consider how these traits would impact their behavior in a reality show setting.

4. **Outline the Plot:** Collaborating with AI, students will outline the plot of their reality show, detailing key events, challenges, and interactions among the characters. They should think about how conflicts might arise and how alliances could form based on the personalities involved.

5. **Write a Script or Episode Summary:** Using their outline and with an LLM, students will write a script or a detailed episode summary that highlights key moments in the reality show, including dialogue snippets, challenges, and character dynamics.

6. **Present Their Reality Show:** Students will create a trailer for the episode with key excerpts from the script or summary, focusing in on intriguing plot points or unexpected twists they've incorporated into the show.

Target Skills:

- Creative Writing
- Character Analysis
- Critical Thinking
- Prompt Engineering

Variations:

- **We've Got History:** Have students create scenarios where characters from opposing historical backgrounds or literary genres must confront their differences, providing opportunities for conflict and resolution.
- **Nielsen'd:** After screening the trailers, the class can vote on their favorite reality show concepts and what made them so compelling or entertaining.
- **Junket:** As an extension, students can create a mock interview with a character from someone else's reality show to find out the real backstory and what they really think about their fellow contestants.

Shirt Shop

In this activity, we'll reimagine the 'headlines' thinking routine to create catchy T-shirt graphics based on a reading passage, book, or historical event. This exercise will help them identify the main idea and collaborate with an LLM to express it creatively.

Procedure:

1. **Select a Reading Passage or Event:** Choose a reading passage, book, or significant historical event for students to analyze. Ensure it has clear themes and messages that can be distilled into a main idea.

2. **Introduce the Headlines Thinking Routine:** Explain the Headlines thinking routine, emphasizing its purpose in capturing the essence of what students have learned. Discuss how headlines summarize key points and draw conclusions about a topic.

3. **Identify the Main Idea:** Have students work individually to identify the main idea of the selected text or event. Then have students pair up to discuss and brainstorm what they believe to be the core message.

4. **Collaborate with AI:** Students will then collaborate with an LLM to generate catchy phrases or slogans that capture the main idea. They can prompt the AI with their identified themes and ask for creative suggestions for T-shirt graphics.

5. **Design T-Shirt Graphics:** Using the generated phrases, students will create their T-shirt graphics. They can sketch designs on paper or use AI image generation tools to visualize their tees.

6. **Present Their Designs:** In a section of the room or online, have students display their T-shirt graphics. If you like, students can vote on the tee they love most.

Target Skills:

- Critical Thinking
- Prompt Engineering
- Collaboration

Variations:

- **It's a Group Thing:** Have students work in larger groups to create a series of T-shirt graphics based on the same text or event, fostering teamwork and diverse perspectives.

- **Mock It:** Students can use graphic design software or AI imaging to create digital mockups of their T-shirt designs.
- **Deeper-est:** Students can present their T-shirt graphics to the class, explaining their design choices, the main idea they captured, and how the AI contributed to their creative process.

Critical Thinking and Problem Solving

These activities emphasize higher-order thinking skills like analysis, evaluation, and problem solving. Students engage with AI to analyze information, form arguments, and make decisions based on evidence.

MacGyver This!

Students solve challenges using limited resources, prompting AI for building instructions, encouraging resourcefulness, problem solving, and following directions.

Procedures:

1. **Assign a Challenge:** Present a MacGyver-like situation for students to solve, such as making a bridge, a weapon, or a method to move items from one place to another.

2. **Gather Materials:** Organize students into small groups (3-4 members) and have them collect 4-8 items from the classroom. Possible materials include paperclips, pencils, paper, blocks, binders, and books.

3. **Prompt AI for Instructions:** Instruct students to ask their AI chatbot how to use their collected materials to create the assigned item. Encourage them to request three different options so they can choose the one they think will work best. They should also generate step-by-step building directions.

4. **Approval and Build:** Have students seek approval for their build from a teacher or peer team before starting the construction process.

5. **Build Away!** Once approved, students can begin building their projects.

Target Skills:

- Following Multistep Directions
- Teamwork and Collaboration
- Prompt Engineering
- Critical Thinking

Variations:

- **Add Visuals:** For students who need extra accommodations, add visual checklists or supports to help them understand the directions and execute the activity.

- **Technology Free:** Using an interactive whiteboard, prompt the AI together to generate three building options. Then have students execute the build process in small groups.

- **Research Element:** Ask students to research real-world examples of their assigned build before deciding on the best version from the AI suggestions.

- **Forgot the Permits:** If you don't have the materials or time to actually build, have students render an image of their creation using an AI image generator.

MacGyver Buildables:

- **Simple Structures:** A tower, a bridge, or a boat
- **Creative Creations:** A robot, an animal, or a vehicle
- **Functional Items:** A toy, a costume (holiday or country specific), a gadget (phone holder, pencil sharpener, etc.)
- **Engineering Challenges:** Rube Goldberg Machine (chain reaction machine), hydraulic lift
- **Simple Machines:** A model of a lever, pulley, inclined plane, wedge, screw, or wheel and axle.
- **Artistic Inventions:** A musical instrument, a game, or an upcycled art piece

Monthly Budget Challenge

Students manage a limited budget, using AI to assist in prioritizing expenses and making financial decisions, developing financial literacy and problem-solving skills.

Procedure:

1. **Create a Bills List:** Generate a list of bills due at the end of the month based on local currency and the ages of your students.
2. **Set a Monthly Budget:** Establish a monthly budget that is less than the total amount due in bills.
3. **Utilize AI for Budgeting:** Have students enlist the help of an AI chatbot to determine which bills to pay and which to skip.
4. **Justify Decisions:** Students will discuss whether they agree with the AI's recommendations and justify their opinions.
5. **Group Comparisons:** Form groups of 2-3 students to compare their results and decide who had the best budgeting ideas.

Target Skills:

- Communication Skills
- Critical-Thinking Skills
- Prompt Engineering

- Problem-Solving Skills

Variations:

- **How Short is Short?:** Have students calculate how much they are short for the month based on their budget and bills.
- **Revise that Budget:** Students can create a revised monthly budget that accounts for their bills, income, and any prior shortfalls.
- **Mo Money:** Students will brainstorm 5 plausible ways to generate enough money to cover all their bills by the end of the month, using an AI chatbot for ideas.
- **Tech-Free:** After developing their short budget, students can prioritize spending in pairs. The class can then use an interactive whiteboard to prompt AI and compare its suggestions to the students' decisions.

Redesigner

Students research and propose improvements to existing products, practicing critical analysis and innovative design thinking.

376

Procedure:

1. **Provide Existing Products:** Present students with an existing product or widget (or a list of several options).

2. **Group Assignments:** Organize students into small groups or peer pairs.

3. **Research Similar Products:** Each group will research 3 similar products or widgets and identify their key features.

4. **Brainstorm Improvements:** In their groups, students will brainstorm and create a list of at least 5 ideas for improvements to the original widget.

5. **Identify Changes:** Students should identify at least one major change or two minor changes to propose, along with predictions on how these changes would improve or potentially detract from the widget's performance.

6. **Prototype Creation:** Students can either draw their prototype or use AI to generate an image of their redesigned product.

7. **Self-Assessment:** Each student will complete a self-assessment based on a rubric to evaluate if their redesign meets the established standards.

8. **Class Presentations:** Students will present their redesigned widget to the class. As peers present, the audience will evaluate the redesign based on

innovation, improvement, and the effectiveness of the pitch.

Target Skills:

- Communication Skills
- Teamwork and Collaboration
- Prompt Engineering
- Problem-Solving
- Critical-Thinking Skills

Variations:

- **I Vant to Be Alone:** Instead of peer pairs or small groups, allow some students to work independently. This can appeal to gifted students, those with divergent learning styles, or unique interests.

Widget Ideas:

- **Technology-Based Widgets:** App icons, website buttons, notification banners, or search bars.
- **Physical Widgets:** Door handles, light switches, pens or pencils, or kitchen utensils.
- **Everyday Objects:** Traffic signs, packaging, clothing, or furniture.

Contract Negotiator

Students analyze contracts for loopholes and benefits, encouraging attention to detail and strategic thinking.

Procedure:

1. **Provide a Contract:** Distribute a contract for students to examine.
2. **Peer Pair Discussion:** Students will share the main points of the contract with a partner, ensuring they understand its key elements.
3. **Identify Loopholes:** Students will look for loopholes in the contract, as well as ways to terminate the agreement early or benefit from it.
4. **Group Collaboration:** Find another group to team up with and compare the loopholes identified by each group.
5. **Class Vote:** Each group will vote on the best loophole from both groups to share with the class.

Target Skills:

- Communication Skills
- Teamwork and Collaboration
- Prompt Engineering
- Problem-Solving Skills

Variations:

- **Can't Negotiate This!** Have students rewrite the contract to make it ironclad, closing all loopholes.
- **Analyze Actual Contracts:** Provide students with real-world contracts from various industries (e.g., rental agreements, employment contracts, service contracts) or create a contract that has key loopholes included.
- **Identify Bias:** Have students analyze the contract for any biases or unfair terms that might favor one party over the other.

Ideas:

Here are a few contracts to use:

- **Online Service Agreement:** A contract for an online service, such as a streaming platform or a cloud storage service.
- **Employment Contract:** A basic employment contract outlining the terms of employment between an employer and an employee.
- **Rental Agreement:** A rental agreement for an apartment or a house.
- **Sales Contract:** A contract for the sale of goods.
- **Non-Disclosure Contract:** An agreement between two parties limiting the information that can be shared or disclosed

Real Photo or AI?

Students try to create hyper realistic images with AI, then analyze the differences between real and AI-generated content, developing visual literacy and critical evaluation skills.

Procedure:

1. **Locate a Real Image:** Students will find a real image online that they would like to replicate.
2. **Generate an AI Image:** Using an AI tool, students will prompt the AI to generate a similar version of the chosen photo.
3. **Share and Compare:** Once students believe their AI-generated image is as realistic as possible, they will share it with a peer partner or small group.
4. **Vote on Real vs. AI:** Within their groups, students will vote on which image they believe is real and which is AI-generated.
5. **Reflect on Realism:** Students will reflect on what aspects made one image appear more realistic than the other and discuss any giveaways that indicated AI generation.
6. **Broader Voting:** The best-performing images will be presented to the larger group for broader voting and discussion.

Target Skills:

- Prompt Engineering
- Critical-Thinking Skills

Variations:

- **Chat It Again!:** Students can take their successful prompt and input it into another image generation model. They will then compare the results against the original image and evaluate the differences.

Talk with History

Students interact with an AI "historical figure" and evaluate the accuracy of its responses, emphasizing research and source validation skills.

Procedure:

1. **Choose Historical Characters:** Have students select from a set of historical figures or experts relevant to the grade level standards.
2. **Prompting AI:** Guide students on how to prompt the AI to respond as the chosen historical figure or topic expert.
3. **Brainstorm Questions:** Students will brainstorm three key questions they would like to ask their selected historical figure based on prior instruction or lesson objectives.
4. **Engage with AI:** Students will prompt the AI with the questions they generated and engage in a dialogue.
5. **Evaluate Responses:** After receiving responses, students will evaluate the AI character's answers as probable, likely, unlikely, or improbable, and find source evidence to support their evaluations.

Target Skills:

- Communication Skills
- Prompt Engineering
- Critical-Thinking Skills

Variations:

- **Debate Me!:** Have students debate with the historical persona they've selected, challenging its views and arguments.
- **Time Travel Much?:** Have students ask modern-day questions or address contemporary issues with the historical figure as if they were transported in time.

It's a Conspiracy!

In this activity, students explore a conspiracy theory and use an LLM to either prove or disprove it. This activity will enhance their critical thinking, research skills, and understanding of how to evaluate claims.

Procedure:

1. **Select a Conspiracy Theory:** You identify a specific conspiracy theory for the class to investigate that aligns with your content or grade level standards. This could be a well-known theory, such as those surrounding historical events or popular culture, or you can make one up!

2. **Ask a Thought Provoking Question:** Have students consider this: 'How can you tell if a conspiracy theory is real or not?' alone or in pairs.

3. **Introduce the Concept of Conspiracy Theories:** Discuss what conspiracy theories are, including their characteristics and why people may be drawn to them. Highlight the psychological aspects and the importance of critical thinking when evaluating such claims.

4. **Group Formation:** Divide the class into small groups. Each group is tasked with either proving or disproving the selected conspiracy theory.

5. **Research Phase:** Groups will use an LLM to gather information. They can prompt the AI with questions related to the conspiracy theory, seeking evidence, counter arguments, and relevant historical context.

6. **Analyze the Information:** Students will review the information they found and consider its relevance and reliability. They will then select their strongest argument.

7. **Present Opposing Ideas:** Students will partner with someone on the other side of their theory. As they present arguments and evidence to each other, they discuss why people could believe both sides.

Target Skills:

- Critical-Thinking Skills
- Research Skills
- Collaboration Skills

Variations:

- **Have you Heard This One?** After students settle the conspiracy theory debate, they can create a new and related conspiracy theory of their own! It should be related to the original one they worked on, and use similar fears, emotions, and tropes to be more convincing.

I Object!

Students use AI to develop arguments for debates, promoting research, persuasive writing, and critical analysis of different viewpoints.

Procedure:

1. **Generate a Debatable Topic:** Ask students a thought-provoking question and use their responses to generate a target debatable topic.
2. **Build Arguments Using AI:** Have students use AI to formulate arguments for both sides of the debate. This includes preparing rebuttals for the main arguments and objections.
3. **Divide and Assign Sides:** Randomly divide students into two groups, assigning one group to the affirmative side and the other to the negative side.
4. **Preparation Time:** Set a timer for 5 minutes. The affirmative side will prepare their case based on the strongest arguments from each student, while the negative side will compile their strongest rebuttals.
5. **Debate Time:**

      a. The affirmative side selects a representative to present their strongest argument for 2 minutes.

      b. The negative team then sends a representative to provide a rebuttal for 2 minutes.

      c. Repeat this cycle 3 times, rotating speakers from each group.

6. **Evaluation:** At the end of the debate, you can act as a judge, or the other students in the class can determine if the case was strong enough to defend against the rebuttal.

Target Skills:

- Communication Skills
- Critical Thinking
- Research Skills
- Problem-Solving Skills

Variations:

- **Deb-AI-te Me!:** Instead of debating with one another, students can debate with AI! They can choose to argue for or against the topic and see how persuasive they can be. Here is a suggested prompt:

- ○ "Let's debate [topic]. I will argue [for/against] it."
- ○ "I believe [topic and side]. Take an opposing view and debate me."

Timelines Visualized

Students create timelines, demonstrating their understanding of historical sequence and cause-and-effect.

Procedure:

1. **Identify a Historical Moment:** Choose a significant moment in history or a key story/book that has a narrative to explore.
2. **Generate a Timeline:** Students will create a timeline of significant moments leading up to and following the chosen event.
3. **Group Collaboration:** Have students meet in small groups or peer pairs to compare their timelines. Encourage discussion and collaboration to enhance their timelines.
4. **Adjust Timelines:** Students can adjust their group timelines as needed based on insights gained during the discussion.
5. **Visualize the Timeline:** Once the final timeline has been created, have groups visualize their timelines by creating AI-generated images or drawings that represent key events.
6. **Display and Review:** Have students display or share their timelines with the class. Allow time for

everyone to review one another's creations and provide feedback.

Target Skills:

- Critical-Thinking Skills
- Prompt Engineering

Variations:

- **Mega-Time:** Merge several groups to create a comprehensive timeline that includes multiple historical events or narratives, allowing for broader discussions and connections between different timelines.
- **Weaving Webs:** At the start of the process, assign different characters or historical figures centered around a specific event. When it's time to Mega-Time (above), create the timeline in a way that gives insight on how individual journeys intersect.

Multiverse

Students alter historical timelines and analyze the potential consequences, practicing hypothetical thinking and understanding the impact of decisions.

Procedure:

1. **Create a Timeline:** Have students create a timeline that includes 10 key moments, numbering them from 1 to 10. These moments can be drawn from history or based on significant events in books.
2. **Identify Significance:** Students will use AI to identify the importance of each moment on their timeline and document its significance in relation to historical or narrative context.
3. **Choose a Divergence Point:** As the teacher, select a number from 1 to 10 (preferably one of the latter numbers). This number will represent the point where the timeline splits into an alternate reality.
4. **Rewrite History:** Students will imagine the moment that the chosen number went in a different direction. They will then rewrite every point on the timeline that follows this divergence, outlining how events would change.

5. **Pair and Share:** Students will pair up to share their new multiverse predictions, discussing the implications of the altered timeline and the decisions made.

Target Skills:

- Communication Skills
- Prompt Engineering
- Critical-Thinking Skills
- Problem-Solving Skills

Variations:

- **It's Too Many:** If 10 points seems like too many, have students make a timeline with just 5 key moments.
- **Can You Picture It?** Have students use AI to generate an image of this newly imagined world in the multiverse to share alongside their predictions.

Outta Time

Students consider the potential impact of futuristic items on historical events, encouraging creative problem solving and historical analysis.

Procedure:

1. **Identify a Significant Moment:** Choose a significant day in history or an important moment in a story that will serve as the focus of the lesson.
2. **Imagine Time Travel:** Have students imagine they can travel back in time to the identified moment and provide one person or character with a single item from the future.
3. **Generate a List of Items:** Students will use AI models to generate a list of 5 futuristic items they could give, along with the rationale behind each choice.
4. **Pair Discussion:** Students will pair up to discuss their lists, sharing their reasoning and any additional insights.
5. **Rate the Lists:** After discussing, students will rate their lists from best to worst based on perceived impact and effectiveness.
6. **Write a Paragraph/Paper:** Working solo, students will write a paragraph or paper (with or

without AI support) explaining why they believe the item they chose is the best option and analyzing the potential impact it could have on the identified day or moment.

Target Skills:

- Communication Skills
- Prompt Engineering
- Critical-Thinking Skills
- Problem-Solving Skills

Variations:

- **Did I Just Make That Up?:** Have students make two time jumps. First, they travel to the year 2050. When they arrive, they are able to take one item back to the significance date or moment. Now, they have to imagine items from the future and their impact on the past.

- **Come Here Often?** Have students choose a *person* to bring back to the significant date or moment you've identified. Just like any good time traveling tale, there are rules here. The person cannot run the chance of running into themselves, so no "I'd bring back the older version of themselves to talk some sense into them…".

I'll Take Two

In this activity, students will use LLMs to generate a compelling sales page for a specific item, service, or product inspired by a reading passage, a recent book, or a historical moment. This activity will enhance their understanding of persuasive writing and marketing strategies.

Procedure:

1. **Select a Product or Service:** Select an item, service, or product related to a recent reading passage, book, or historical event.

2. **Introduction to Sales Pages:** Have students work in small groups to identify the elements of a successful sales page, including:
   - Attention-grabbing headlines
   - Engaging taglines
   - Persuasive sales copy
   - Clear calls to action

3. **Share Examples:** Students can share examples of effective sales pages with the class or peer partners to analyze what makes them successful.

4. **Brainstorming Headlines and Taglines:** Students will work with LLMs in small groups

or solo to brainstorm catchy headlines and taglines for their chosen product. Encourage them to think about their target audience and what drives them to buy.

5. **Generate Sales Copy:** Students will use an LLM to help draft their sales copy. Guide them to ask for suggestions on how to present features, benefits, and compelling reasons to purchase.

6. **Write the Sales Page:** Using the AI-generated content as a foundation, students will compose a full sales page that includes:
    - A captivating headline
    - An engaging tagline
    - Persuasive sales copy that highlights unique selling points
    - A clear call to action encouraging readers to make a purchase

7. **Peer Review and Feedback:** Students will pair up to review each other's sales pages, providing constructive feedback on clarity, persuasiveness, and overall appeal.

8. **Present to the Class:** Each student or group will present their sales page to the class, discussing their chosen product, the creative

process, and how they incorporated persuasive techniques.

Target Skills:

- Critical-Thinking Skills
- Research Skills
- Communication Skills
- Prompt Engineering Skills

Variations:

- **It's On!:** Organize a friendly competition where students vote on the most persuasive sales page. Consider categories such as "Most Creative," "Best Headline," and "Overall Best Sales Page."

Communication and Collaboration

This category focuses on activities that promote communication and collaboration skills, but also integrating AI to enhance interactions.

Social Skills Role PL-AI

Students practice social interactions through role-playing or role pl-AI-ing, using AI to generate scripts and scenarios, developing communication and teamwork skills.

Procedure:

1. **Post a Scenario:** Present a social situation for students to practice, such as a job interview, a community interaction, or a general social scenario.
2. **Divide into Groups:** Organize students into small groups of 2-3 members.
3. **Generate Scripts with AI:** Students will use an AI chatbot to generate a script for their role-play. They should prompt the AI with the number of characters, the scenario, and some loose plot points to guide the conversation.
4. **Self-Assign Roles:** Within their groups, students will self-assign roles based on the generated script and practice their parts.
5. **Perform Role-Plays:** Groups will perform their role-plays in front of their peers, showcasing their interpretations of the scenarios.

6. **Provide Feedback:** After each performance, encourage students to give constructive feedback on what worked well and areas for improvement.

Target Skills:

- Communication Skills
- Teamwork and Collaboration
- Prompt Engineering
- Problem-Solving Skills

Variations:

- **Communication Systems (AAC):** For students who use alternative communication systems or voice output devices, preprogram them as part of the process and work to cue and prompt the student to participate.
- **Technology-Free:** Using an interactive whiteboard, prompt the AI together to generate a role play script. Then have students practice and perform.
- **Partial Scripting:** Generate a script together, but leave out the ending. Have students improve a new ending or write one before they perform it.
- **Real-World Scenarios in the News:** Search for a recent faux pas or social stumble and use that as

the foundation for your role play, encouraging students to explore real-life implications.

- **Build Empathy:** Have students create characters with diverse backgrounds, perspectives, and challenges.

Permission Slip?

Students create travel guides using AI for multimedia elements, practicing research, organization, and presentation skills.

Procedure:

1. Identify a specific location related to your lesson objectives OR let students to choose any location.
2. Students create a guide, including historical facts, cultural insights, and location-based activities.
3. Have students incorporate multimedia elements (video, audio, slideshows) based on their guides.

Target Skills:

- Communication Skills
- Prompt Engineering
- Research Skills

Variations:

1. **Epic Tour:** Preset 5-10 locations and have students pick based on preference. Working in small groups, they will work to create a presentation that allows for all your students to go on a world tour. Just remember to bring your permission slip.

I Teach, Therefore I Am

Students teach AI about a topic they have learned, reinforcing their understanding through explanation and addressing AI-generated questions.

Procedure:

1. **Learn a Topic:** After students learn about a specific topic or finish reading a passage/book, they will partner with AI to engage in a teaching exercise.

2. **Prompt AI:** Students will prompt AI with the following statement:
   - "I just learned about [topic]. Let me explain it to you, and you can ask me questions about it. Here is what I learned…"

3. **Set a Time Limit:** Students will engage in this teaching exercise for a set amount of time, allowing them to articulate their understanding and respond to AI-generated questions.

4. **Generate Questions:** When the time is over, students will compile a list of questions they still have on the topic or questions that the AI posed that they didn't know the answers to.

5. **Question Parking Lot:** Students will add these questions to a "question parking lot" in the classroom or their personal enquiry list to research during their PEACE time.

Target Skills:

- Communication Skills
- Prompt Engineering
- Critical-Thinking Skills

Variations:

- **Did You Know That?:** Students can partner with a peer or small group after step 3 to share the questions that AI had and explain the answer to the group. This allows all students to benefit from concept reinforcement!

Preach!

In this activity, students will take a historical text or a character's speech and transform it into a modern-day speech or song, enhancing their understanding of historical context and creative expression.

Procedure:

1. **Choose a Historical Text or Speech:** You can pick a historical speech or a significant speech from a character in literature. This could include speeches from figures like Martin Luther King Jr., Susan B. Anthony, or characters from novels and plays.

2. **Analyze the Speech:** Students will analyze the chosen text to identify its main themes, messages, and emotional appeals. They should consider the historical context and the impact of the speech at the time it was delivered. This can be done with a LLM or in small groups.

3. **Transform the Speech:** Solo, students will rewrite the speech or transform it into a modern-day speech or song. They should work with the LLM to capture the essence of the original while making it relatable to contemporary audiences.

4. **Include Original Quotes or Segments:** As part of their transformation, students will include three quotes or segments from the original speech.
5. **Present Their Work:** Students will present their modern speeches or songs to the class, sharing the significance of their chosen quotes.

Target Skills:

- Communication Skills
- Critical Thinking
- Creative Expression
- Historical Analysis

Variations:

- **Let's Collab:** Have students work in small groups to choose a single speech and create a collective modern adaptation, allowing for diverse perspectives and collaborative creativity.
- **It's a Movement!:** Let students incorporate multimedia elements (e.g., music, visuals) into their remake.
- **Run that Back:** After presenting, have students discuss how the original speech compared to the modern adaptation. Was the same message relayed, even with a change in language, style, and relevance.

- **Talk About a Revolution:** Choose a speech that addresses social issues and discuss how the modern adaptation could spark action or community awareness today.

Going Viral!

In this activity, students will create a viral social media post based on a character or historical figure. It should be catchy and students will have to cite text evidence or historical evidence to support their content.

Procedure:

1. **Select a Character or Historical Figure:** Choose a character from literature or a significant historical figure that aligns with content and grade-level standards. You can present several options so students can choose.

2. **Discuss Characteristics of Viral Content:** Have students research what elements make content go 'viral' on social media, such as humor, relatability, emotional appeal, and current relevance. Students can pair with a partner to discuss examples of viral posts they were able to find.

3. **Research Phase:** Students can use an LLM to research text evidence or historical evidence about their chosen character or figure. They should focus on key traits, significant events, quotes, or interesting facts that could capture an audience's attention.

4. **Brainstorm Ideas:** Students will brainstorm ideas for their social media posts. Encourage them to think about what platform they are targeting (e.g., Instagram, Twitter, TikTok) and how they can adapt their message accordingly. Then, they will meet with a small group or partner to evaluate ideas and give some feedback.

5. **Create the Viral Post:** Students will create their social media post, which may include:
   - A catchy caption or tagline and body copy.
   - Relevant images or graphics (can be drawn, digitally created, or sourced).
   - Hashtags to increase visibility.
   - Any additional elements such as polls, questions, or challenges that encourage engagement and support the character or figure's intent.

6. **Justify with Evidence:** Students will write a brief rationale explaining how their post is supported by text or historical evidence.

7. **Present and Share:** Students will present their viral posts to the class or small groups. They can explain their creative process, the evidence they used, and why they believe their post would be successful on social media.

Target Skills:

- Creative Thinking Skills
- Research Skills
- Communication Skills

Variations:

- **One Million Likes:** After presentations, let the class vote for the "most viral" post. Discuss what elements contributed to its likable appeal.
- **Reposting:** Have students create variations of their posts for different social media platforms, discussing how messaging and presentation change across platforms.
- **Life Before the Gram:** Have students discuss how historical figures might use social media today and consider how their messages would have to be adapted to modern audiences.

Chapter 8

Bringing PEACE to Your Classroom

I still remember the first time I watched my students really get something... like, truly see them piece it together in a way that made sense to them. Not to me, not to a curriculum standard, but to the kid sitting right there. For a teacher, that's gold, right? That's the moment we all signed up for. And I know, it doesn't happen every day. Between assessments, behavior management, and the endless sea of admin tasks, those moments can feel buried. But what if that feeling didn't have to be rare? What if creating a classroom where every student has the chance to wonder, question, and create could actually be your everyday reality?

Here's the thing: AI integration in the classroom doesn't mean you're surrendering to the machines or sacrificing meaningful learning for the sake of shiny tech. It's quite the opposite. By harnessing AI thoughtfully, through the PEACE Framework, you're creating a learning environment that's dynamic, student-centered, and perfectly suited for a future that none of us can fully predict. Think about it: You've got enquiry at the core, AI at your side, and students who are building skills they'll need whether they go on to be doctors, designers, or data scientists.

I know there are always reasons *not* to try something new. Maybe you're thinking, "My kids already struggle with the basics," or "How can I add one more layer to my already full plate?" Fair points. But here's the beauty of the PEACE Framework: It's a structure, yes, but one that's flexible. You're not starting from scratch; you're infusing what you already do with tools and approaches that spark curiosity and build real-world skills. It's not about adding more to your plate - it's about using what's on it more effectively.

Remember those early days we talked about? Week one, just dipping your toes in with a familiar lesson, a few simple questions, and maybe a test drive with ChatGPT or Claude. From there, you built your rhythm, took chances, reflected, and tried again. It wasn't perfect. But it's in the process (the back and forth, the successes and slip-ups) that you truly begin to see the magic of enquiry in action. And when students see their voices and choices reflected in your teaching, the buy-in is real. Suddenly, it's not just a lesson; it's their lesson. And that's powerful.

By now, you know this isn't about replacing the teacher. It's about enhancing what you already do so well: Building a classroom where students don't just memorize but internalize, where they don't just complete tasks but engage with them. And, yes, where they don't just listen but question. The PEACE Framework, paired with AI, can take you there. It's not a gimmick or a shortcut; it's a pathway that empowers you to teach for the future, not just in it.

So, here's my question for you as you turn the last page: What do you want your students to remember? The tests, the worksheets, the flashcards? Or do you want them to remember the questions they asked, the answers they found, and the thrill of creating something meaningful? The choice is yours, and you're already well on your way.

Thank you for taking this journey. May your classroom be a place of discovery, growth, and a whole lot of "aha" moments. And may you always have a little PEACE along the way.

Glossary

AAC: Augmentative and Alternative Communication tools and methods help individuals with difficulty speaking or communicating, often used in special education to support students with disabilities.

AI (Artificial Intelligence): A field of computer science that creates intelligent systems capable of tasks like learning, problem-solving, and decision-making, requiring a transformative shift in teaching and learning.

AI Literacy: The ability to understand, evaluate, and use AI effectively and ethically, recognizing its benefits, limitations, and implications.

Algorithm: A set of instructions AI systems follow to process information and perform tasks, forming the foundation of how AI learns and makes decisions.

Authentic Assessment: Assessments that require students to apply knowledge in real-world contexts, such as through projects, presentations, or portfolios.

Automation: Using AI to perform tasks automatically, like grading or generating materials, freeing teachers for more meaningful interactions with students.

Bias: Prejudice or favoritism in data or algorithms that can lead to unfair outcomes or perpetuate stereotypes.

BID Routine: A question protocol (**B**rainstorm, **I**dentify, and **D**raft) designed to refine student enquiry and focus questions.

Black Mirror Scenario: A dystopian situation where technology, often seemingly beneficial, has unforeseen and negative consequences on society and individuals. Inspired by the British TV series *Black Mirror*.

Chatbot: A computer program that simulates conversation with humans, often used in education for personalized tutoring, feedback, and support. Can also be used to reference interactions with LLMs.

CLEAR Check: A process for refining AI interactions, helping students assess if responses are **C**omplete, **L**ogical, and well-**E**xplained, and then **A**djusting the output or **R**ewording the query for better results.

Ed Tech (Educational Technology): Tools and platforms used to support and enhance teaching and learning, from interactive whiteboards to AI-powered systems.

Elements: Layers added to questions to deepen complexity and encourage higher-order thinking, such as **temporal** (time-related), **spatial** (location-based), **comparative** (evaluating differences or similarities), and other dimensions that deepen enquiry and move beyond simple recall.

Exit Ticket: A short activity at the end of a lesson to assess understanding and gather feedback, adaptable for AI-enhanced learning.

Executive Functioning: Mental skills for planning, organizing, managing time, and regulating emotions, with AI tools offering structured support and reminders.

Figma: An online tool for collaborative brainstorming, visual thinking, and organizing ideas, useful in various stages of the PEACE Framework.

George Orwell: A renowned author whose works, such as *1984*, explore the ethical implications of technology, often referenced in discussions about AI's societal impact.

Google-able Question: A question that can be quickly answered with a basic internet search, typically requiring low-level recall.

Hallucination: AI-generated information that is factually inaccurate or unsupported by training data, highlighting the importance of critical evaluation.

Hal 3000: A fictional sentient computer from *2001: A Space Odyssey*, symbolizing the ethical and societal dilemmas of AI.

Inquiry-Based Learning (IBL): A student-centered approach emphasizing exploration, questioning, and discovery to build understanding through active engagement, foundational to the PEACE Framework. Also referred to in this book as enquiry-based learning.

Instructional Pedagogy: The art and science of teaching, encompassing the theories, methods, and approaches used to facilitate learning. Shifting to the Intelligence Age, it embraces AI as a partner in the learning process, moving away from traditional, teacher-centered models.

Iterative: A process involving repeated cycles of refinement and improvement, essential in both AI interactions and student learning.

Large Language Model (LLM): AI systems trained on extensive text data to generate and understand human language, such as ChatGPT.

Multiverse: A concept that suggests the existence of multiple universes outside of our own. These "parallel universes" or "alternate realities," differ in small ways (like tiny changes in events or outcomes) or can be completely different (with different physical laws, dimensions, or even forms of life).

Open-Ended Question: A question with no single correct answer, encouraging diverse responses, allowing for wonder, and pushing towards deeper enquiry.

Padlet: An online tool for collaborative multimedia content creation, useful for sharing research and feedback as part of enquiry-based learning.

Personalized Learning: Tailoring educational experiences to individual student needs, supported by AI's ability to adapt instruction and feedback.

PEACE Framework: A five-step inquiry-based learning system designed for an AI-enhanced classroom, including: **P**rovoke, **E**nquire, **A**nalyze, **C**reate, and **E**ngage. The PEACE Framework provides a structure for using AI to enhance curiosity, creativity, and critical thinking.

Progressive Check-in: An informal, ongoing assessment method to monitor student understanding and provide timely feedback during a project or learning process.

Prompt/Prompting: The process of asking a generative AI tool to perform a task or provide information, often requiring clarity and specificity for optimal results.

Prompt Engineering: The skill of crafting precise and effective prompts or questions to elicit desired responses from AI tools. It involves understanding how AI models work, using specific language and formatting, and iteratively refining prompts to achieve the best results.

Question Literacy: The ability to formulate, refine, and evolve questions that lead to meaningful discovery and deeper understanding.

Question Stems: Sentence starters or prompts that guide students in formulating questions, often used as scaffolding tools.

Question Storm/Question Storming: A brainstorming activity where students generate a wide range of questions on a topic to explore different angles and possibilities.

ReBID Routine: An iterative extension of the BID routine where students **R**eview and **E**valuate their original question, then **B**rainstorm, **I**dentify, and **D**raft (BID) an improved version.

Ruler Moment: A metaphorical term describing a moment when student curiosity is piqued, leading to genuine wonder and a desire to explore further.

SAFE Check: A process for evaluating AI responses, focusing on the **S**ource, **A**ccuracy, **F**airness, and **E**vidence to ensure credibility and balance in the information.

Scaffolding: Temporary support to help students complete tasks they couldn't independently, possibly provided by AI, through hints or feedback.

Scantron: A brand of standardized test forms and scoring machines often associated with traditional, multiple-choice assessments that typically required number 2 pencils to bubble in test answers.

SHARP Filter: A system for crafting high-quality enquiry questions, where SHARP stands for **S**pecific, **H**igh-level, **A**ctionable, **R**elevant, and **P**owerful. It encourages students to move beyond basic, Google-able questions toward those that spark deeper thinking and investigation.

Think Aloud: A teaching strategy where the teacher models their thought process out loud to demonstrate effective questioning or problem-solving techniques.

Thinking Routines: Structured strategies or processes that guide thinking and learning, such as brainstorming, concept mapping, and questioning techniques.

Thought-Provoking Question: A question that requires deeper analysis, synthesis, or evaluation, encouraging higher-order thinking and exploration.

Traditional Assessments: Standardized tests, worksheets, and other methods that rely on memorization and recall.

About the Author

Ayo Jones is an educational innovator transforming special education through technology and teaching. With a Master's in Instructional Design and Technology, she has trained thousands of teachers and administrators at the Regional Education Service Centers in Texas and also as a featured contributor on the Texas Education Agency's Special Education Support website, impacting tens of thousands of students across Texas and beyond.

Her innovative approach to integrating technology into special education has pioneered one-to-one device programs and created new models for teaching students with complex needs. But it was her bold decision to sell everything and move her family to Ghana, West Africa (a journey captured in the HBO Max series *Coming From America*) that truly exemplified her commitment to educational transformation. In Ghana, she volunteers at the community's only autism school, expanding her mission to support students with disabilities globally.

Through her platform Noodle Nook, her podcast, and her active presence on LinkedIn and YouTube, Ayo continues to share resources and strategies with educators worldwide. When she's not revolutionizing special education or developing new teaching tools, you'll find her pursuing her passion for creative problem solving through LEGO building - an interest that mirrors her approach to education: Building better futures, one piece at a time.

Learn from Ayo at www.PeaceFramework.com
Link with Ayo at www.PeaceFramework.com/linkedin
Watch Ayo at www.PeaceFramework.com/youtube